checkpoint

chemistry

Jasmine

checkp■int

Endorsed by
University of Cambridge
International Examinations

chemistry

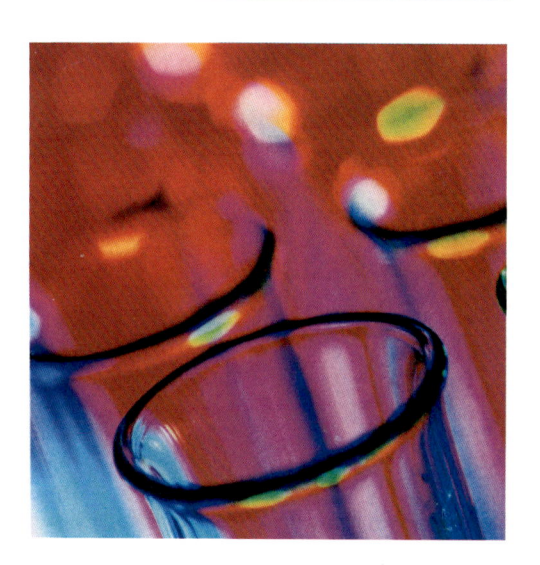

Peter D Riley

HODDER EDUCATION
PART OF HACHETTE LIVRE UK

Titles in this series
Checkpoint Biology Pupil's Book — ISBN 978 0 7195 8067 3
Checkpoint Biology Teacher's Resource Book — ISBN 978 0 7195 8068 0
Checkpoint Chemistry Pupil's Book — ISBN 978 0 7195 8065 9
Checkpoint Chemistry Teacher's Resource Book — ISBN 978 0 7195 8066 6
Checkpoint Physics Pupil's Book — ISBN 978 0 7195 8069 7
Checkpoint Physics Teacher's Resource Book — ISBN 978 0 7195 8070 3

To Len and Ann

Orders: please contact Bookpoint Ltd, 130 Milton Park, Abingdon, Oxon OX14 4SB. Tel: (44) 01235 827720. Fax: (44) 01235 400454. Lines are open from 9.00–5.00, Monday to Saturday, with a 24-hour message answering service. Visit our websites www.hoddereducation.co.uk and www.hoddersamplepages.com

© Peter Riley 2005
First published in 2005
by Hodder Murray, an imprint of Hodder Education,
part of Hachette Livre UK
338 Euston Road
London NW1 3BH

Impression number   10 9 8 7 6 5
Year   2010 2009 2008 2007

All rights reserved. Apart from any use permitted under UK copyright law, no part of this publication may be reproduced or transmitted in any material (including photocopying or storing in any medium by electronic means and whether or not transiently or incidentally to some other use of this publication) without the written permission of the Publisher except in accordance with the provisions of the Copyright, Designs and Patents Act 1988 or under the terms of a licence issued by the Copyright Licensing Agency Limited, Saffron House, 6–10 Kirby Street, London EC1N 8TS.

Cover design by John Townson/Creation
Illustrations by Mike Humphries, Linden Artists
Typeset in 12/14 Garamond by Pantek Arts Ltd, Maidstone, Kent
Printed in Dubai

A catalogue record for this title is available from the British Library

ISBN: 978 0 7195 8065 9

Contents

Preface		vi
Acknowledgements		ix
Introduction		1
Chapter 1	Acids and bases	20
Chapter 2	Physical and chemical changes	29
Chapter 3	Investigating everyday materials	45
Chapter 4	Particle theory	55
Chapter 5	Mixtures and separating techniques	62
Chapter 6	Atoms and elements	75
Chapter 7	Further reactions	85
Chapter 8	Compounds and mixtures	99
Chapter 9	Metals and non-metals	109
Chapter 10	Corrosion	133
Chapter 11	Patterns of reactivity	137
Chapter 12	Preparing salts	141
Chapter 13	Exothermic and endothermic reactions	144
Chapter 14	Rates of reaction	152
Chapter 15	The periodic table	161
Appendix		171
Glossary		174
Index		178

Preface

To the pupil

Chemistry is the scientific study of matter and materials. These are substances such as air, water and rock, which make our world. Matter is made from tiny structures called atoms that join together in many different ways to make millions of different kinds of chemicals.

The first atoms formed at the beginning of the Universe. At that time a huge explosion called the Big Bang took place. The first atoms formed two gases – hydrogen and helium. Over thousands of millions of years since then many other atoms were made in the stars. When the stars faded out or exploded in a supernova, these atoms spread through space and formed dust. In time, one cloud of dust formed the Solar System and everything that is in it. This means that all the materials you will study in your chemistry course once formed in the Universe billions of years ago. It also means that the materials from which you are made formed then too.

Our knowledge of chemistry has developed from the observations, investigations and ideas of many people over a long period of time. Today this knowledge is increasing rapidly as there are more chemists – people who study matter and materials – than ever before.

In the past, few people other than scientists were informed about the latest discoveries. Today, through newspapers, television and the internet, everyone can learn about the latest discoveries on a wide range of chemical topics, from making new materials for exploring the oceans or space, and developing new medicines and fuels, to finding ways to recycle the materials we use to make them available for future generations.

Checkpoint Chemistry covers the requirements of your examinations in a way that I hope will help you understand how observations, investigations and ideas have led to the scientific facts we use today. The questions are set to help you extract information from what you read and see, and to help you think more deeply about each chapter in this book. Some questions are set so you can discuss your ideas with others and develop a point of view on different scientific issues. This should help you in the future when new scientific issues, which are as yet unknown, affect your life.

PREFACE

The scientific activities of thinking up ideas to test and carrying out investigations are enjoyed so much by many people that they take up a career in science. Perhaps *Checkpoint Chemistry* may help you to take up a career in science too.

To the teacher

Checkpoint Chemistry has been developed from *Chemistry Now! 11–14* second edition to specially cover the requirements of the University of Cambridge Checkpoint tests and other equivalent junior secondary science courses. It has the following aims:

- to help pupils become more scientifically literate by encouraging them to examine the information in the text and illustrations in order to answer questions about it in a variety of ways; for example, 'For discussion' questions may be used in work on science and citizenship;
- to present science as a human activity by considering the development of scientific ideas from the earliest times to the present day;
- to examine applications of scientific knowledge and issues that arise from them.

This *Pupil's Book* begins with an introduction which briefly reviews the development of science, and in particular chemistry, throughout the world; it then moves on to consider the scientific method and its application. As many consequences of scientific work raise issues, the introduction concludes by addressing an issue with a chemical base and shows strategies that pupils can use to consider the problems and try and resolve them. These strategies can be used in the rest of the book in discussion activities where issues with a chemical base arise. The introduction also contains material from Chapter 1 of *Chemistry Now! 11–14* second edition, which introduces the pupils to equipment used in chemistry and advice on how to behave in a laboratory.

The chapters are arranged in the order of the topics in the Cambridge Checkpoint Chemistry Scheme of Work. Chapters 1–5 address topics for Year 1, Chapters 6–10 address topics for Year 2, and Chapters 11–15 address topics for Year 3. The pupils are introduced to chemical symbols and word equations in appropriate sections of the book. On page 171 there is an appendix

PREFACE

in which the construction of chemical formulae and symbol equations are explained. The appendix ends by showing how state symbols can be used in symbol equations.

There is an extensive glossary at the back of the book, which includes all the words designated in each topic as essential to the pupil's scientific vocabulary.

This *Pupil's Book* is supported by a *Teacher's Resource Book* that provides answers to all the questions in the Pupil's Book – those that occur in the body of the chapter and those that occur as end-of-chapter questions.

The *Teacher's Resource Book* also provides end-of-chapter tests, which can be used for extra assessment, and actual questions from past Checkpoint tests. There is a range of practical activities for integration with the work in each chapter – providing opportunities for pupils to develop their skills in scientific investigation.

Acknowledgements

Cover, p.iii Tek Image/Science Photo Library; **p.1** Science Photo Library/Jerry Lodriguss; **p.2** Still Pictures/Adrian Arbib; **p.3** *t* Science Photo Library/Erich Schrempp, *b* Science & Society Picture Library/Science Museum; **p.6** Andrew Lambert; **p.7** *both* Andrew Lambert; **p.8** Andrew Lambert; **p.9** *t* Andrew Lambert, *b* Science Photo Library/Damien Lovegrove; **p.10** *all* Andrew Lambert; **p.11** Science Museum/Science & Society Picture Library; **p.12** Andrew Lambert; **p.17** *t* Science Photo Library/David Taylor, *b* Science Photo Library/Louise Lockley/CSIRO; **p.18** Ecoscene/Sally Morgan; **p.20** *l* John Townson/Creation, *c* Phil Chapman, *r* John Townson/Creation; **p.21** Andrew Lambert; **p.22** *both* John Townson/Creation; **p.23** *t* John Townson/Creation, *c* Pete Atkinson, *b* Ardea/Ken Lucas; **p.24** Andrew Lambert; **p.26** *t* John Townson/Creation, *bl* Andrew Lambert, *br* Ecoscene/Chinch Gryniewicz; **p.27** Natural Visions/Heather Angel; **p.30** *l* John Townson/Creation, *r* Science Photo Library/Peter Menzel; **p.31** Powerstock; **p.32** John Townson/Creation; **p.34** Alamy/Doug Young Travel; **p.35** *t* Corbis/David Ball, *b* Bob Battersby; **p.36** *l* Science Photo Library/Adrienne Hart-Davis, *r* Science Photo Library/Martyn F. Chillmaid; **p.37** Alamy/Dex Image; **p.39** *t* Corbis/Tim Mosenfelder, *b* Andrew Lambert; **p.40** Alamy/Phil Degginger; **p.41** Alamy/Robert Slade; **p.43** *l* Science Photo Library/Andrew Lambert Photography, *c* Andrew Lambert, *r* Andrew Lambert; **p.45** Corbis/Lawson Wood; **p.46** *t* Corbis/Arne Hodalic, *c* Corbis/Paul A. Souders, *bl* Alamy/Phil Broom, *br* Bob Battersby; **p.47** *t* Corbis/Michael S. Yamashita, *b* Zefa/Masterfile/Burazin; **p.48** *both* Bob Battersby, **p.49** *t* Science Photo Library/Jim Reed, *b* Alamy/Bruce Coleman Inc.; **p.50** *t* Corbis/Owaki-Kulla, *b* Alamy/View Pictures Ltd.; **p.51** *t* Corbis/Earl & Nazima Kowall, *b* Alamy/Adrian Sherratt; **p56.** Natural Visions/Heather Angel; **p.57** *t* John Townson/Creation, *b* Alamy/David Lyons; **p.60** John Townson/Creation; **p.61** *all* Andrew Lambert; **p.62** NHPA/David Woodfall; **p.63** Science Photo Library/Biophoto Associates; **p.64** *all* Andrew Lambert; **p.68** John Townson/Creation; **p.70** *both* Andrew Lambert; **p.73** *both* John Townson/Creation; **p.78** Andrew Lambert; **p.79** Wellcome Library, London; **p.81** Science Photo Library/Alfred Pasieka; **p.83** Rex Features/Juhan Kuus; **p.86** *all* John Townson/Creation; **p.87** *t* Geoscience Features, *b* John Townson/Creation; **p.88** *t* John Townson/Creation, *b* Empics; **p.89** Science Photo Library/David Taylor; **p.90** *t* Science Photo Library/Ian Boddy, *b* John Townson/Creation; **p.91** *t* John Townson/Creation, *b* Alamy/Jeff Smith; **p.92** Royal DSM

ACKNOWLEDGEMENTS

N.V. **p.93** *all* Andrew Lambert; **p.95** Corbis/Paul A. Souders; **p.96** *t* Corbis/Paul A. Souders, *b* Mary Evans Picture Library; **p.97** Corbis/Jose Manuel Sanchis Calvete; **p.99** *all* John Townson/Creation; **p.100** John Townson/Creation; **p.101** Rex Features/ISOPRESS; **p.102** Andrew Lambert; **p.103** Andrew Lambert; **p.105** Ecoscene/Adrian Morgan; **p.111** *both* Geoscience Features Picture Library; **p.112** John Townson/Creation; **p.113** *t* Pirelli Cables Ltd., *b* Ancient Art and Architecture Collection/Ronald Sheridan; **p.116** Robert Harding Picture Library; **p.117** Jaguar Cars Ltd.; **p.121** Andrew Lambert; **p.122** Ecoscene/Sally Morgan; **p.123** *t* Ecoscene/Wayne Lawler, *b* Ecoscene/Chinch Gryniewicz; **p.124** Panos Pictures/Marc Schlossman; **p.125** Still Pictures/Elizabeth Herbert; **p.126** US Department of Energy/Science Photo Library; **p.130** Andrew Lambert; **p.133** Martyn Pitt/MRP Photography; **p.134** *t* Bob Battersby, *b* John Townson/Creation; **p.135** Still Pictures/UNEP; **p.136** *t* Corbis/Wolfgang Kaehler, *b* Alamy/Diomedia; **p.137** *t* Science Photo Library/Lawrence Migdale, *bl* Andrew Lambert, *br* Andrew Lambert; **p.138** *both* Andrew Lambert; **p.144** Corbis/John & Lisa Merrill; **p.148** Corbis/Wally McNamee; **p.149** *t* John Townson/Creation, *b* Lorna Ainger; **p.151** Geoscience Features; **p.152** *t* Getty Images, *b* Bob Battersby; **p.158** Science Photo Library/Astrid & Hanns-Frieder Michler; **p.164** Science Photo Library; **p.165** *tl* Geoscience Features Picture Library, *tr* Geoscience Features Picture Library; **p.166** Paul Brierley Photo Library; **p.167** *l* © ILFORD Imaging, *r* © ILFORD Imaging; **p.168** *t* Rex Features/Mark Brewer, *b* Science Photo Library/David Parker; **p.169** Landmark Media/Mark Beltran; **p.171** *both* Science Photo Library/Adam Hart-Davis; **p.172** Science Photo Library/Will and Deni McIntyre.

t = top, *b* = bottom, *l* = left, *r* = right, *c* = centre

Every effort has been made to contact copyright holders but if any have been inadvertently overlooked the Publishers will be pleased to make the necessary arrangements at the earliest opportunity.

Introduction

Figure 1 Stars – chemical factories in the sky.

When you look up at stars in the night sky, you are looking at hundreds of chemical factories. In each star a chemical called hydrogen is being squashed so hard by the star's gravity that it changes to another chemical called helium. When this happens, a huge amount of energy is released as heat and light. The production of chemicals does not end with helium. Many more chemicals including carbon, oxygen and iron are made. Eventually a star runs out of hydrogen and begins to 'die'. If it is small it may simply fade away and the chemicals it has made drift off into space. If the star is large it may explode as a huge supernova. The energy released in this explosion is so great that more chemicals like gold or silver are formed. They rush out into space with the rest of the materials from the exploding star.

In time clouds of hydrogen and helium remaining in space join with chemicals from 'dead' stars and swirl around. The hydrogen and helium form a new star at the centre of the swirling mass and the chemicals form spheres around it – planets and their moons. Wherever you are on Earth now, you are surrounded by chemicals made in stars. The oxygen you breathe and the carbon in your clothes formed from stars that existed long ago. Even the chemicals which make your bones and blood formed in the same way.

> **For discussion**
> In some science fiction stories, terms such as 'people of the stars' or 'star children' are used. Is there any link between stars and people? Explain your answer.

The first chemists

The very first human species used materials to help them survive. They clothed themselves in animal skins to keep warm, arranged branches into shelters to keep dry and sharpened stones to make axes. They used physical processes such as cutting to help them. The first

INTRODUCTION

people to use a chemical reaction to help them were the people who used fire. It is known that people about three-quarters of a million years ago in France had used fire. A hearth – a place where fire is made regularly – has been found and dated to that time.

As with every chemical reaction, fire making needed some preparation. Wood had to be dried. Some wood had to be made into small dry shavings to act as tinder. Two other pieces of wood had to be selected to rub together to produce heat. Once enough heat was generated next to the tinder, it burst into flame as the carbon in the wood combined with the oxygen in the air in a chemical reaction.

Figure 2 Making fire using a fire drill.

The heat energy from the fire was used to bring about other chemical reactions. Probably the first chemical reactions to be brought about by fire were those involved in the cooking of food. About 7000 years ago people in Turkey used fire to bring about chemical reactions which turned clay into pottery. Four thousand years ago the Egyptians used fire to release copper from its ore. Five hundred years later, the Hittites in Turkey discovered that iron could be released from its ore by heating the ore with charcoal.

INTRODUCTION

Figure 3 Chinese firecrackers.

From alchemy to chemistry

At about the same time, metal workers in a civilisation called Mesopotamia, a region between the rivers Tigris and Euphrates, looked for a way of turning ordinary metals into gold and silver. They also looked for a way of making a medicine that could make people live for ever. The ideas and processes that they carried out in their investigations became known as alchemy. As the people traded with countries to their west and east the practice of alchemy spread along the trade routes into Arabia, Africa, India and China.

Alchemists never found the substance that they called the philosopher's stone which would change ordinary metals into gold and silver. They also never found the elixir of life – the medicine which would allow people to live for ever. They did, however, make some discoveries about chemicals.

In China, alchemists discovered that saltpetre (chemical name potassium nitrate), sulphur and carbon could be mixed together to make gunpowder. They used it to make fireworks and added a range of other substances such as tiny pieces of iron to make sparkles and cotton fibres to make a violet colour.

The Chinese also introduced new materials. They invented paper by mashing plant fibres and then letting them dry to form sheets. Much of the paper was made from the bark of the mulberry tree. The paper was used to make clothing, armour, shoes and mosquito nets in addition to its use as a writing material. The Chinese also tapped the sap from the lacquer tree and made a varnish which was the first plastic. They used the lacquer to coat furniture, boxes, kitchen utensils such as ladles, and even used it in making coffins.

When the Muslim countries developed in Arabia alchemy was greatly studied. Probably the greatest of the Muslim alchemists was Jabir Ibn Haiyan who was also known as Gerber. He spent most of his time developing and improving ways of making chemical investigations such as making crystals or distilling liquids. He discovered different kinds of acids, worked on the development of steel and divided substances into groups such as metals and non-metals. His work eventually led other scientists away from alchemy to the study of chemistry and he is known as the 'father of chemistry'. He died in 803 CE (the year AD 1).

Figure 4 The alembic was invented by Gerber to distil liquids.

INTRODUCTION

> **For discussion**
> Construct a time line showing the development of fire, the first making of pottery, the first extraction of copper and iron, the beginning of alchemy and the beginning of chemistry. What scale should you use? Find out about the dates when other materials and chemicals were discovered and add them to your time line.

Alchemy continued for another 800 years after Gerber's time but others began to follow his example and spend more time investigating substances than looking for ways to become wealthy and live for ever. In 1661 Robert Boyle, an Irish scientist, wrote a book called *The Sceptical Chymist*. He used the word sceptical in the title because he no longer believed in some of the ancient ways of the alchemists. Boyle dropped the 'al' from alchemist to make chemist (or as he spelled it *chymist*) because of his beliefs in experimentation, observation, recording and using experimental evidence to explain how substances are made and how they behave. From that time chemistry became a science.

Making a new substance

Alchemists in the past were fascinated by making new substances just like chemists are today. Here is a simple experiment in which you can make a plastic from milk and vinegar. Follow the steps and record your observations as you go along, as Boyle encouraged all scientists to do.

1. Measure out 300 ml of milk into a beaker and let your teacher warm it for you. The milk should not get so hot that it boils.
2. Add 15 ml of vinegar to the milk and stir the two liquids together.
3. Leave the mixture to cool for 15 minutes.
4. Place a kitchen sieve over a second beaker and pour the mixture into the sieve.
5. Squeeze any solid substance in the sieve to remove more liquid.
6. Tip the solid in the sieve onto a paper towel.
7. Examine the substance two hours later, a day later and two days later.

> **For discussion**
> What kinds of chemical research do you think are being carried out today? Check your ideas by finding out what chemical research is being done at the university nearest to your school. Log onto its website and look for the chemistry department web pages.

INTRODUCTION

Issues

Some of the facts that have been discovered through scientific investigation can be used to help us survive – through helping us produce new materials for clothing and housing. Scientific facts can also help when dealing with many issues in everyday life. When people have different ideas about solving a problem, the problem becomes an issue. A very simple issue could be the state of your room. You may think that it is tidy while other members of your family want you to tidy it up. How do you resolve this issue? Here is an imaginary example of an issue in which science facts can be helpful. Read about it – then try the exercises in dealing with the issue.

Kando is a village on a river by the edge of a forest. There are farms close by the village where most of the people work. They do not make much money from their work but the forest provides them with fuel and the river provides them with fish. There is no electricity supply to the village and the only way to reach the neighbouring town is along a very bumpy road which floods in the wet season.

One day geologists visited the area and discovered that under the forests and the farms was a chemical that is needed for making metals for jet engines.

Later the owner of a mining company visited the village. He explained to the villagers about the importance of the chemical in the ground beneath their feet and how he would like to set up a mine to remove it. The mining company owner said that large areas of forest and farmland would have to be cleared away but there would be plenty of work and money for the people who worked in the mine.

Some people in the village thought that the mine was a good idea and would improve life for the villagers. Other people in the village thought that the mine was a bad idea and would make living conditions worse.

> For discussion
> Read the story about Kando to your friends and present your cost–benefit analysis. Ask for your friends' views and discuss them. Is the issue resolved satisfactorily?

One way of considering an issue like this is to make a cost–benefit analysis. This can be simply done by making a list of advantages and disadvantages about setting up a mine in the village and comparing them.

For example, one advantage of setting up the mine would be greater air travel for people around the world. A disadvantage might be pollution of the river caused by mining waste.

Make your cost–benefit analysis about setting up the mine in the village by writing down as many advantages and disadvantages as you can.

INTRODUCTION

What is chemistry?

Chemistry is the study of the structure of substances and how they change. In the following pages you can learn some of the practical skills you need to begin your study of chemistry.

Measuring quantities

For many investigations, the quantities of substances taking part in a chemical reaction need to be known and also the quantities of the substances that are produced.

Measuring the volume of a liquid

The volume of a liquid can be found by pouring it into a measuring cylinder and reading the scale. A measuring cylinder can also be used to prepare a specified volume of a liquid by pouring in an amount of the liquid, then either topping it up or pouring some out, until the required volume is present in the cylinder.

A burette is used to deliver a required volume of liquid, but it cannot be used to find the volume of a liquid in the same way that a measuring cylinder can.

The scales on measuring cylinders and burettes indicate the volume in either millilitres (ml) or cubic centimetres (cm^3). A millilitre is a thousandth of a litre and is the unit used in measuring out liquids that are sold in bottles and cans. It is also used to indicate the quantities of liquids that are used in recipes. In scientific work the unit cm^3 is used and $1\,cm^3 = 1\,ml$.

Measuring cylinder

A liquid is poured into the cylinder and the volume is read from the scale on the side. The surface of the liquid curves upwards at the point where it touches the inside of the cylinder. This curvature is called the meniscus. To read the volume of a liquid accurately, the base of the measuring cylinder must be placed on a flat surface and the eye must be level with the surface of the liquid in the middle of the cylinder (see Figure 5). The volume of liquid in the cylinder in Figure 5 is $35\,cm^3$.

1 How would reading the volume of a liquid from the top of the meniscus make the reading inaccurate?

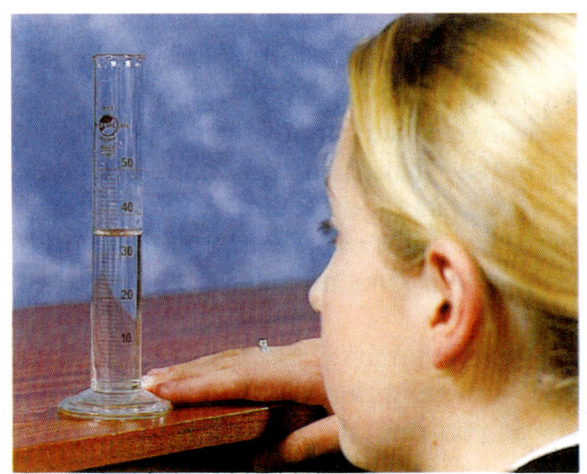

Figure 5 Reading the volume of liquid in a measuring cylinder.

6

INTRODUCTION

The meniscus of mercury is unusual in that it curves downwards. You would never measure mercury in a measuring cylinder (it is too dangerous as mercury vapour escaping from the surface could be inhaled and poison the body), but you can see its meniscus in a mercury-in-glass thermometer.

Burette

The liquid is poured into the top and the burette is filled up to the zero mark on the scale. Liquid is drawn from the burette by opening the tap at the bottom (see Figure 6). The amount of liquid that has been released from the burette in Figure 6 is 11 cm^3.

Measuring the volume of a gas

The volume of a gas may be measured using a syringe with a scale marked on it. The scale may measure in millilitres or cubic centimetres. As the gas is produced it passes along a tube into the syringe and pushes out the plunger. The volume collected in the syringe can be measured by reading the scale at the place where the plunger comes to rest (see Figure 7).

Figure 6 A burette scale reads from the top down.

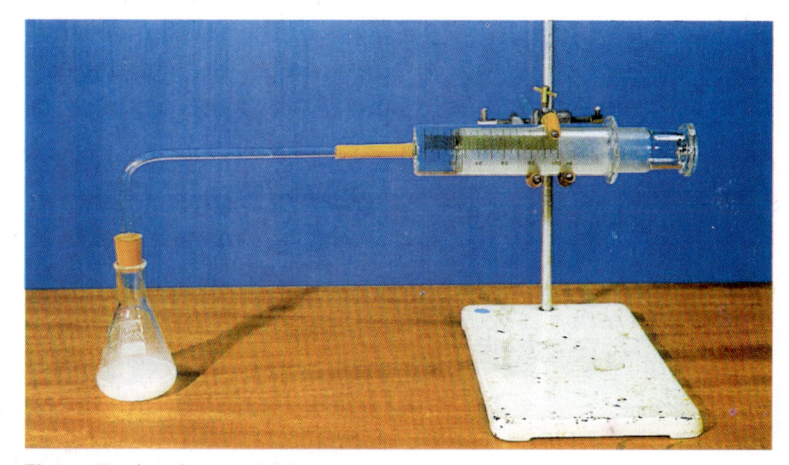

Figure 7 A syringe containing a gas.

Measuring the mass of a solid or a liquid

The mass of a substance can be found by using a top loading balance. This piece of equipment is very sensitive and must be treated with great care at all times. Loads for weighing must be put onto, and removed from, the pan carefully.

The mass of the substance in Figure 8 (on page 8) is found by reading the main number display and then reading the two decimal places from the second number display. The mass of the substance in Figure 8 is 17.24 g.

INTRODUCTION

Figure 8 A top loading balance.

The top loading balance measures mass in grams (g) but it also has a mechanism which allows larger masses to be measured in kilograms (kg). For most laboratory work the balance is used to measure small masses in grams.

If a substance such as a liquid is to be weighed in a beaker, the mass of the beaker must first be found. The mass of the substance and the beaker is then found, and the mass of the substance is calculated by subtracting the mass of the beaker from this total. For example:

$$\text{mass of beaker} = 50.00\,g$$
$$\text{mass of beaker + substance} = 120.00\,g$$
$$\text{mass of substance} = 120.00 - 50.00 = 70.00\,g$$

Some balances have a tare. This mechanism allows substances to be weighed out without having to make a calculation. It is used in the following way: the empty beaker is placed on the top pan and its mass is displayed. The tare is then used by turning or pushing a control on the side of the balance. This action brings the reading on the balance back to zero, even though the beaker is still on the pan. The substance can then be placed in the beaker and its mass is read directly from the display.

Measuring temperature

The thermometer is used to measure temperature (see Figure 9). It is a glass tube, with a small container called a bulb at one end. In the container is a liquid which expands or contracts as the temperature changes. The liquid may be mercury or coloured alcohol.

INTRODUCTION

A small amount of the liquid forms a thread in the thermometer tube. As the temperature rises the liquid expands and the thread in the tube increases in length. When the temperature falls the liquid contracts and the thread decreases in length. The length of the thread is measured on a scale on the tube. The Celsius scale is used on most laboratory thermometers. This scale is used to measure the temperature in degrees Celsius (°C).

The end of the thermometer that does not have the bulb may be capped with a piece of plastic, designed to stop the thermometer rolling along the bench and falling onto the floor. Some people fail to tell the difference between the cap and the bulb when they first use a thermometer and use it the wrong way up!

When the temperature of a liquid is to be measured, the bulb of the thermometer should be put into the liquid and the movement of the mercury or coloured alcohol in the thermometer observed. When the expansion or contraction is finished, the temperature can be read from the scale. While the temperature is being read the bulb of the thermometer must be kept immersed in the liquid. If it is removed the temperature of the air will be recorded.

A clinical thermometer like that shown in Figure 10 used to be used to take the temperature of a person. The thermometer is left in place for a few minutes before the temperature is read. It can be removed from the person for reading because it has a narrow bend in the tube which prevents the mercury, which has expanded along the scale, from returning to the bulb. When the thermometer has been read, the mercury is moved back to the bulb by shaking the thermometer.

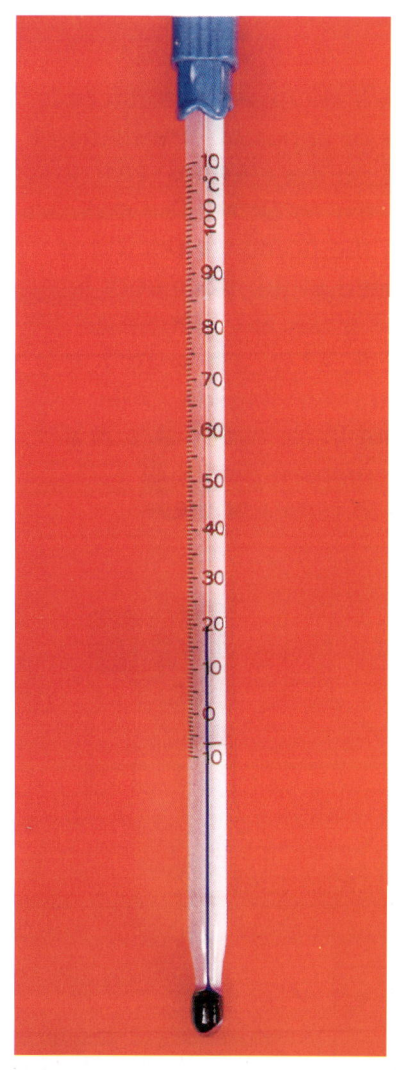

Figure 9 A thermometer.

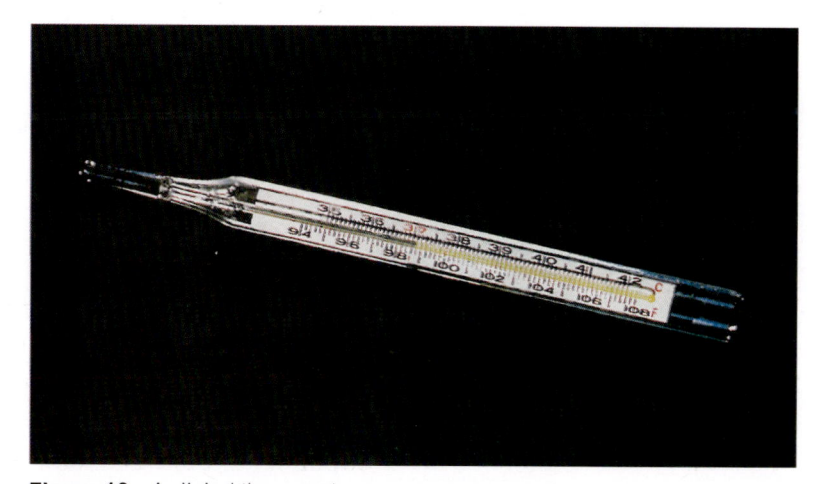

Figure 10 A clinical thermometer.

2 What temperature does the thermometer in Figure 9 show?

3 Someone is asked to take the temperature of a liquid. They put the bulb of the thermometer in the liquid for a few minutes, then take it out to read it. Will their reading be accurate? Explain your answer.

4 What advice would you give to someone who was taking the temperature of a liquid?

INTRODUCTION

Apparatus

The equipment that is used in a chemistry laboratory is called apparatus. Many pieces of apparatus are made of glass because it is transparent, so the chemical reactions are easy to see. Glass is also easy to clean. The ordinary glass used in objects found in the home breaks if it is heated. The glass apparatus used in the laboratory is usually made from borosilicate glass, also known as Pyrex. This glass does not break when it is heated. It is also used to make kitchen glassware like casserole dishes, which can be safely put in an oven to cook a meal. Figure 11 shows some common pieces of apparatus and diagrams used to represent them.

Figure 11

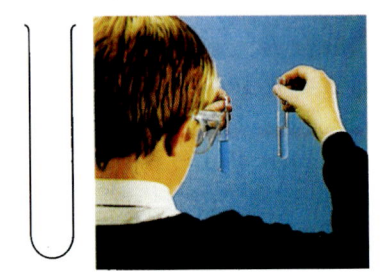

Test-tubes

Round bottomed flask

Flat bottomed flask

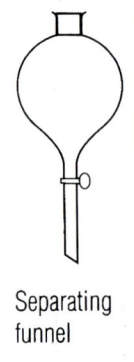

Separating funnel

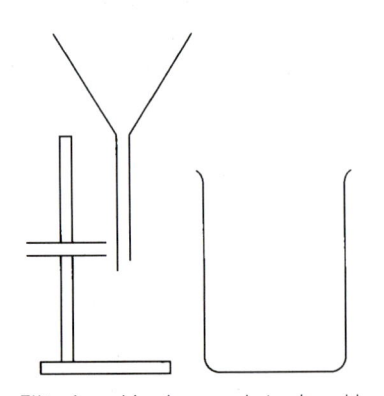

Filter funnel in clamp and stand, and beaker

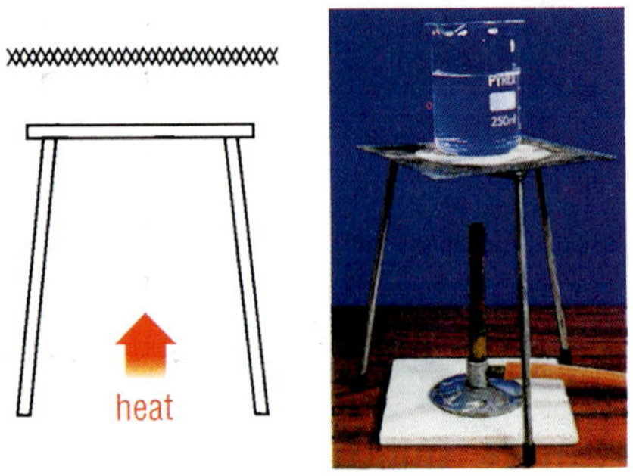

Bunsen burner, tripod and gauze

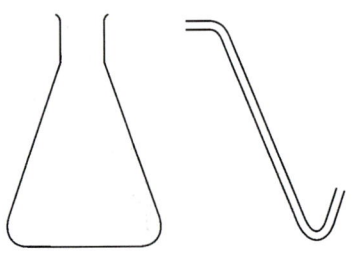

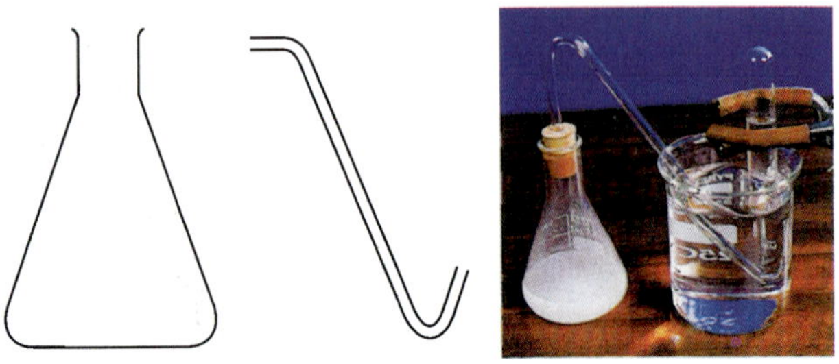

Conical flask with delivery tube

INTRODUCTION

Bunsen burner

Robert Bunsen (1811–1899) was a German scientist. He made many investigations and his work included the invention of a battery and developing a way of identifying substances from the flames they produced. This method has been developed to identify substances in stars.

Bunsen is best known for the Bunsen burner, although he did not, in fact, invent it. However, he used it so widely in his investigations that other scientists began to use it too. Today it is used in laboratories throughout the world to give a strong steady source of heat without smoke.

By using the burner, Bunsen and a colleague discovered two new elements (see Table 6.1 page 75).

1 In which century did Bunsen live?
2 How old was he when he died?
3 Why is the burner named after him?
4 What are the advantages of using a burner, compared with a fire or a candle?
5 Who was Bunsen's colleague, and what were the elements they discovered?
6 How old was Bunsen when he discovered the new elements?

Figure A Robert Bunsen.

When a record of an experiment is being made, a diagram of how the apparatus was set up is included. Each piece of apparatus can be represented diagrammatically so the way the apparatus was set up can be clearly seen.

5 What are the three pieces of apparatus represented by the diagrams in Figure 12?

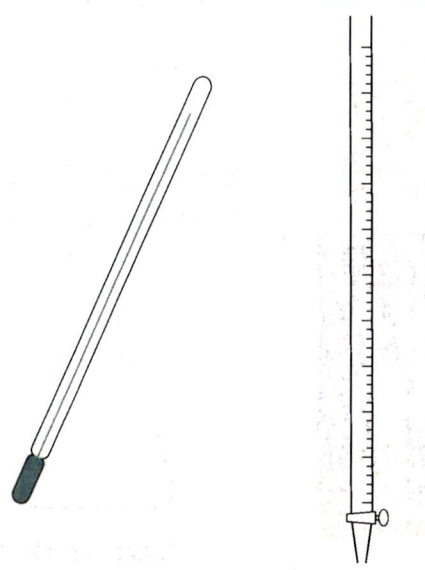

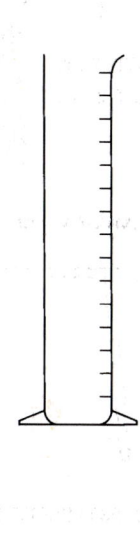

Figure 12

INTRODUCTION

6 Figure 13 shows how some pieces of apparatus were set up in an experiment. Draw a diagram of them and label each one.
7 What is the volume of liquid in the measuring cylinder in Figure 14 a)?
8 How would you measure 10 ml of liquid out of a full burette?
9 How much liquid has been removed from each burette in Figure 14 b)?
10 What is the volume of gas in the syringe in Figure 14 c)?
11 What is the mass of the beaker on the balance in Figure 14 d)?

Figure 13 Apparatus for a simple experiment.

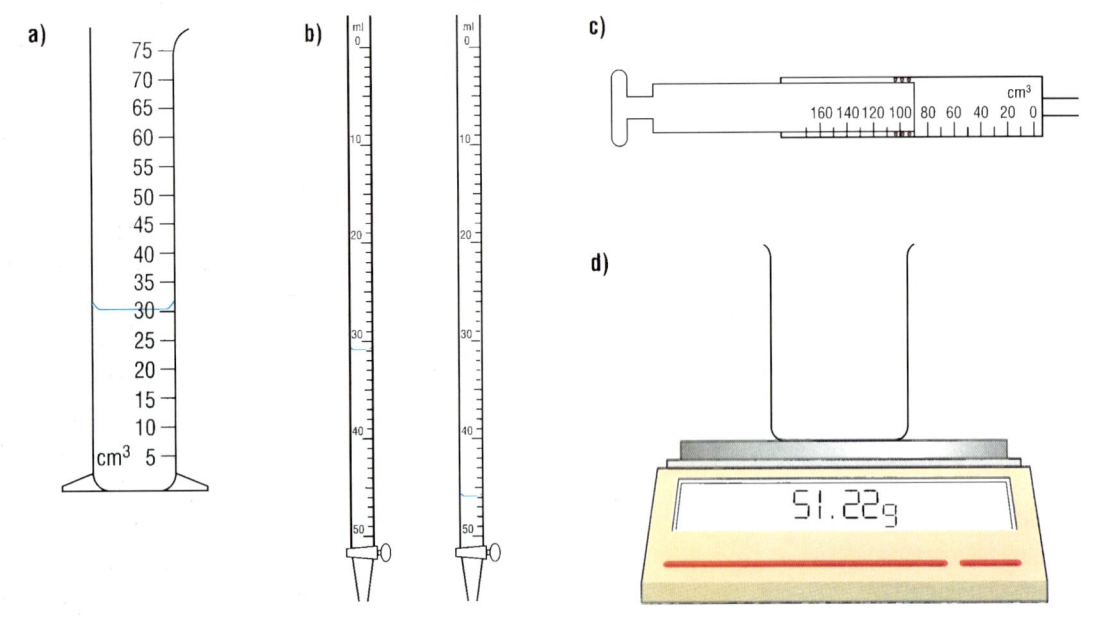

Figure 14

More complicated apparatus

Some pieces of apparatus are complicated. For example, the Liebig condenser is used to convert steam into water. Figure 15 shows the Liebig condenser set up with other pieces of apparatus to carry out distillation (see also page 73). In this apparatus set-up, also note how bungs with tubes passing through them are represented diagrammatically. The condenser is named after Justus von Liebig (1803–1873), a famous chemist of the 19th Century.

INTRODUCTION

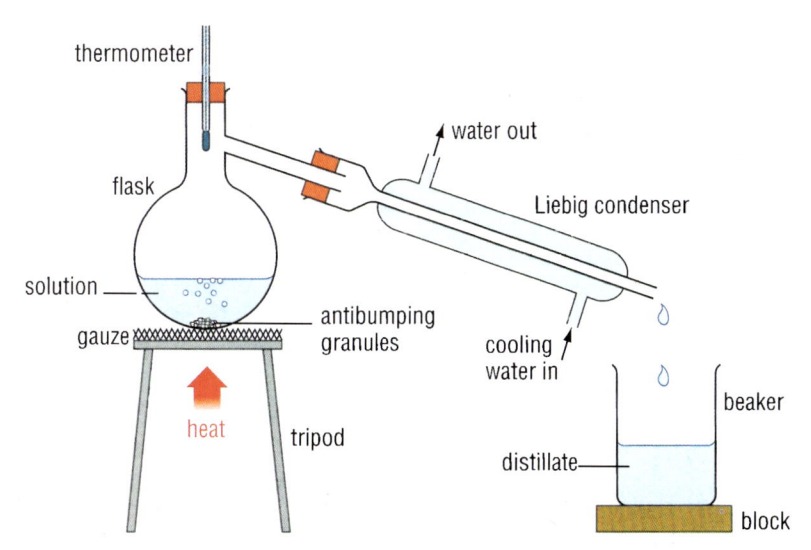

Figure 15 Apparatus for distillation with a Liebig condenser.

Laboratory rules

School laboratories are busy places. There may be about 30 people doing investigations in a laboratory at the same time. They may be using gas, water, electricity, a wide range of glass apparatus and some hazardous chemicals. Despite the large amount of activity, there are fewer accidents in laboratories than in most other parts of a school. The reason for this is that when people work in laboratories, they generally take great care to follow the advice of the teacher and the rules pinned to the laboratory wall.

Laboratory rules can be set out in many ways, but should cover the same good advice. Here is an example.

Entering and leaving the laboratory
- Do not run into or out of the laboratory.
- Make sure that school bags are stored safely.
- Put stools under the bench when not in use.
- Leave the bench-top clean and dry.

INTRODUCTION

General behaviour
- Do not run in the laboratory.
- Do not eat or drink in the laboratory.
- Work quietly.

Preparing to do practical work
- Tie back long hair and if lab coats are available wear them, buttoned up.
- Wear safety spectacles when anything is to be heated or if any hazardous chemicals are to be used.

During experiments
- Never point a test-tube containing chemicals at anyone, and do not examine the contents by looking down the tube.
- Tell your teacher about any breakage or spillage at once. If you are at all unsure of the practical work, check with your teacher that you are following the correct procedure.
- Only carry out investigations approved by your teacher, and use the gas, water and electricity supplies sensibly.

for discussion

What are the reasons for each of the laboratory rules? What other rules could you add?

Figure 16 Good laboratory practice.

INTRODUCTION

Figure 17 Bad laboratory practice.

What are they doing wrong?

Paul ran into the chemistry laboratory because he was keen to do an experiment. He did not see the stool that was sticking out from under the bench and fell over it. He grabbed hold of the bench to stop his fall but his fingers ran into a pool of liquid that had been left on the bench-top and his hand slid, lost its grip and he fell to the floor.

The rest of the class had sat down by the time Paul had picked himself up and put his bag down in the middle of the space between the benches. As Jenny came back from the teacher's bench with a lighted taper for her Bunsen burner, she stumbled against Paul's bag. Her long hair swayed forwards into the taper flame. She jerked her head back and only the tips of a few strands of hair were singed.

Brian had lit his Bunsen burner and was holding a test-tube of liquid over the flame. He was eager to look down the test-tube and brushed aside the safety spectacles that Andrew was holding out for him. The liquid boiled quickly; a few drops shot out of the test-tube and just missed Brian's face.

'Look at that!' he exclaimed, and pointed the test-tube at Paul so he could see too. Jane put down the apple that she was secretly eating to see what Brian and Paul were doing. When she picked it up again, she did not notice the dark, sticky substance clinging to it that had come from the bench-top. She quickly put her apple back into her bag, as the teacher approached to check her experiment.

(continued)

INTRODUCTION

'Did Mrs Jones say to put the apparatus this way round or that way round?' asked Jenny, when the teacher had gone away.

'I don't know. I was too busy unsticking my apple from the bottom of my bag,' replied Jane. 'It looks all right like that. Light the Bunsen burner.'

'That's not right!' shouted Angela. Her loud voice made Brian jump and he dropped his test-tube. Mrs Jones looked round at Angela for a moment, but went off to stop Paul picking up the broken glass with his fingers.

'It should be like ours,' continued Angela in a quieter voice. 'Mrs Jones says it is OK.'

'But Paul's isn't like that,' cried Jenny.

'No,' whispered Paul. 'I'm making up my own experiment. If I light this paper in the sink and put this wire behind that dripping tap and just press this switch then...'

1 List the things that the pupils are doing wrong in this story.

2 What did the pupils who were in the laboratory before this class do wrong?

12 What do you understand by the words:
   a) corrosive,
   b) irritant,
   c) flammable,
   d) radioactive,
   e) toxic?

Warning signs

Like all sciences, chemistry is a practical subject but some of the substances that are used are dangerous if not handled properly. The containers of these substances are labelled with a warning symbol such as those shown in Figure 18.

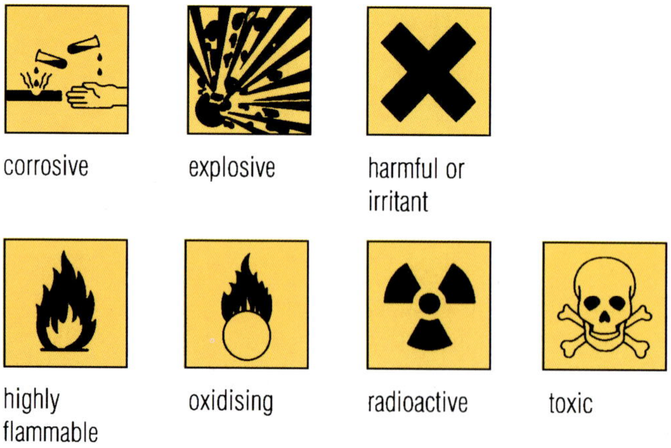

corrosive   explosive   harmful or irritant

highly flammable   oxidising   radioactive   toxic

Figure 18 Warning symbols.

From school laboratory to chemical plant

The laboratories in schools and colleges where chemistry is taught are called teaching laboratories. In these laboratories most of the apparatus is simple, like the pieces shown in Figure 11 on page 10. Some apparatus used for advanced work may be more complicated and is fitted together by ground-glass joints (see Figure 19).

INTRODUCTION

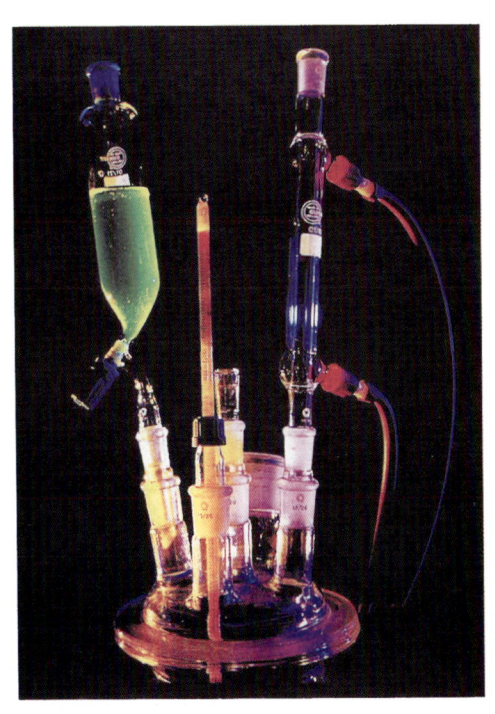

Figure 19 Advanced chemical glassware connected by ground-glass joints.

The most complicated assemblies of apparatus are found in research laboratories where new chemical processes are investigated and new materials are developed (see Figure 20).

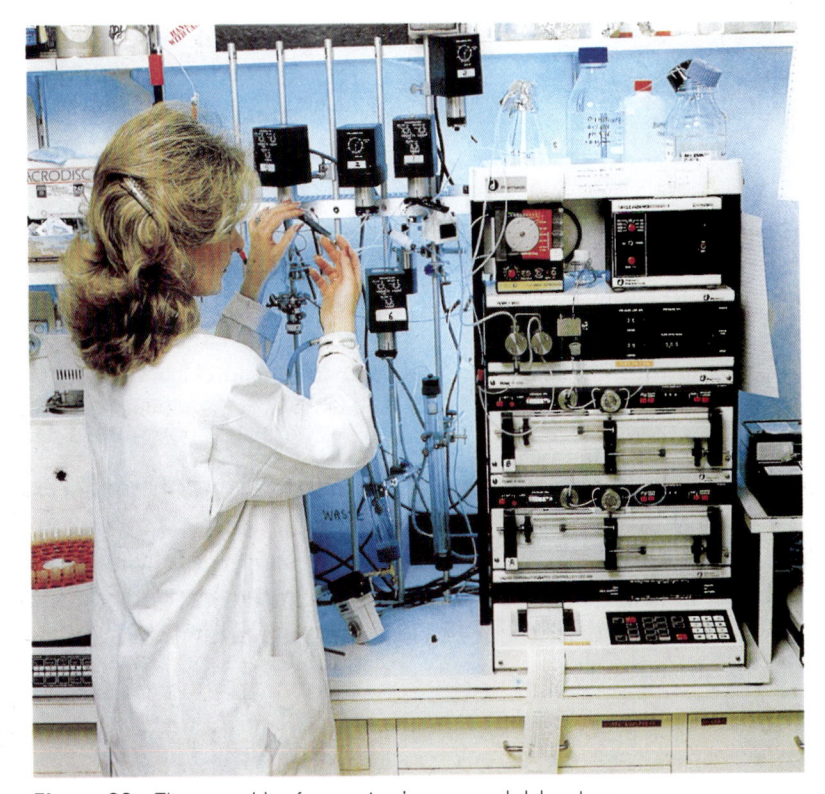

Figure 20 The assembly of apparatus in a research laboratory.

INTRODUCTION

Only small amounts of chemicals are used and made in the apparatus in research laboratories. Later, if it is thought that the process can be used to make large amounts of a material cheaply, a larger version of the apparatus is made and tested to see whether the process continues to work safely. If the larger version is found to be safe then a full-size version of the apparatus, now called a chemical plant, is built. Here, large amounts of chemicals are used to make large amounts of useful materials (see Figure 21).

Figure 21 A chemical plant that produces sulphuric acid.

Chemical reactions and equations

Chemists use equations to describe what happens in a chemical reaction. The equations save time and space and provide the essential information in an easy-to-read form. The simplest equations are word equations.

The substances that take part in a reaction are called the reactants. The substances that form as a result of the chemical reaction are called the products:

$$\text{reactants} \rightarrow \text{products}$$

In an equation the reactants are written on the left hand side and the products on the right hand side. If two or more reactants or products are featured in the equation they are linked together by plus (+) signs:

$$\text{reactant A} + \text{reactant B} \rightarrow \text{product C} + \text{product D}$$

An arrow points from the reactants to the products. Most reactions are not reversible and there is only one arrow. Some reactions are reversible (they can go in either direction) and a special arrow sign is used that points in both directions:

$$A + B \rightleftharpoons C + D$$

13 What is the difference between a product and a reactant in a chemical reaction?

14 How can you tell from the equation if the reaction is reversible or not?

♦ SUMMARY ♦

Fire is probably the first chemical reaction used by humans (*see page 2*).

Alchemists in many countries investigated substances without using the scientific method (*see page 2*).

Chemistry developed from alchemy (*see page 3*).

Science can sometimes be used to resolve issues about how people live (*see page 5*).

Special apparatus is used to measure the volume, mass and temperature of a substance (*see pages 6–9*).

Laboratory apparatus can be represented in diagrams (*see page 10*).

Rules need to be followed to be safe in investigations in the laboratory (*see page 13*).

Warning signs are used on the containers of dangerous substances (*see page 16*).

The work done in school laboratories can be applied in chemical research and the chemical industry (*see page 16*).

Chemists use equations to describe what happens in a chemical reaction (*see page 18*).

End of chapter question

What advice can you give to someone to help them to:
a) work safely in a chemical laboratory,
b) take careful readings in experiments?

1 Acids and bases

Acids

Most people think of acids as corrosive liquids which fizz when they come into contact with solids and burn when they touch the skin. This description is true for many acids and when they are being transported the container holding them has the hazard symbol shown in Figure 1.1.

corrosive **Figure 1.1** The hazard symbol for a corrosive substance.

Some acids are not corrosive and are found in our food. They give some foods their sour taste. This property gave acids their name. The word acid comes from the Latin word *acidus* meaning sour.

Many acids are found in living things. Tables 1.1 and 1.2 show some acids found in plants and animals.

Table 1.1 Acids found in plants.

Acid	Plant origin
citric acid	orange and lemon juice
tartaric acid	grapes
ascorbic acid	vitamin C in citrus fruits and blackcurrants
methanoic acid	nettle sting

Table 1.2 Acids found in animals.

Acid	Animal origin
hydrochloric acid	human stomach
lactic acid	muscles during vigorous exercise
uric acid	urine, excretory product from DNA in food
methanoic acid	ant sting

Figure 1.2 Animals and plants that produce acid.

20

ACIDS AND BASES

The acid in vinegar

Ethanoic acid is found in vinegar and is produced as wine becomes sour. The wine contains ethanol produced by fermentation, and also has some oxygen dissolved in it from the air. Over a period of time, the oxygen reacts with the ethanol and converts it to ethanoic acid. This is an oxidation reaction and the reaction happens more quickly if the wine bottle is left uncorked.

1 Why does wine go sour faster if the cork is removed from the bottle?

Organic acids and mineral acids

The acids produced by plants and animals (with the exception of hydrochloric acid) are known as organic acids. Ethanoic acid is an organic acid and was the first to be used in experiments. Over the period AD 750–1600 the mineral acids were discovered by alchemists. The first mineral acid to be discovered was nitric acid. It was used to separate silver and gold. When the acid was applied to a mixture of the two metals it dissolved the silver but not the gold. Later, sulphuric acid and then hydrochloric acid were discovered. These mineral acids are much stronger (see page 24) than ethanoic acid and allow more chemical reactions to be made. The use of these acids led to many chemical discoveries.

2 How do you think the terms:
   a) organic acids, and
   b) mineral acids came to be used?

3 Acids in the laboratory are stored in labelled bottles as shown in Figure 1.3.
   a) Which acids are dilute and which are concentrated?
   b) How is a dilute solution different from a concentrated one?

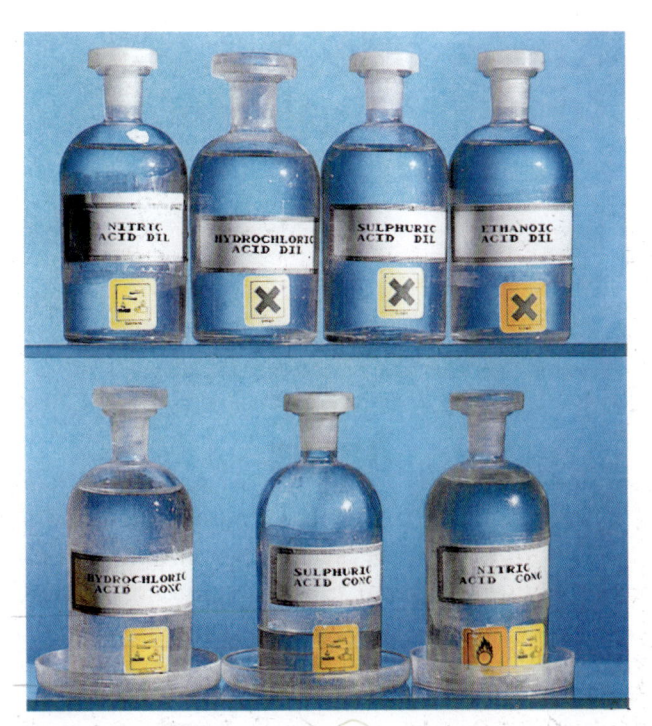

Figure 1.3 Bottles of dilute and concentrated acids.

ACIDS AND BASES

A model volcano

In the past you may have made a model volcano. To do this you may have added a tablespoon of baking soda to an empty plastic drink bottle and then built a mound of sand around the bottle so that it looked like a conical volcano. Finally you may have added red dye to half a cup of vinegar, then poured the vinegar into the bottle. Moments later a red froth would have emerged from the top of the bottle and flowed down the cone of sand, like lava flowing down a volcano (see Figure 1.4). Although the model looks impressive it does not illustrate how lava is formed but it does show the power of a chemical reaction. Vinegar contains an acid but if you were to test the mixture in the bottle for acidity (see page 24) you may not find any. The chemical reaction has neutralised the acid. It is called a neutralisation reaction. A substance which neutralises an acid, like the baking soda, is called a base.

Figure 1.4 The ingredients (left) for making a model volcano (right).

Bases

As bases neutralise acids they are sometimes described as having properties which are opposite to acids. Bases are metal oxides, hydroxides, carbonates and hydrogencarbonates.

Some bases are soluble in water. They are called alkalis. Sodium hydroxide and potassium hydroxide are examples of alkalis that are used in laboratories. When they dissolve they form solutions (see page 64).

ACIDS AND BASES

4 Which of the following substances are bases – copper chloride, sodium hydroxide, calcium carbonate, magnesium sulphate, copper oxide, lead nitrate, sodium hydrogencarbonate?

A concentrated solution of an alkali is corrosive and can burn the skin. The same hazard symbol as the one used for acids (see Figure 1.1) is used on containers of alkalis when they are transported.

Even dilute solutions of alkali such as dilute sodium hydroxide solution react with fat on the surface of the skin and change it into substances found in soap. Many household cleaners (see Figure 1.5) that are used for cleaning metal, floors and ovens contain alkalis and must be handled with great care.

Figure 1.5 Alkalis used in the home.

Detecting acids and alkalis

Some substances change colour when an acid or an alkali is added to them. Litmus is a substance which is extracted from a living organism called lichen. In chemistry it is used as a solution or is absorbed onto paper strips. Litmus solution is purple but it turns red when it comes into contact with an acid. Litmus paper for testing for acids is blue. The paper turns red when it is dipped in acid or a drop of acid is put on it. When an alkali comes into contact with purple litmus solution the solution turns blue. Litmus paper used for testing for an alkali is red. When red litmus paper comes into contact with an alkali it turns blue.

Hydrangeas have pink flowers when they are grown in a soil containing lime (an alkali) and blue flowers when grown in a lime-free soil. The colour of the flowers can be used to assess the alkalinity of the soil.

Figure 1.6 Pink and blue hydrangeas.

ACIDS AND BASES

5 Why are bases sometimes described as the opposite of acids?

6 How are acids and bases similar?

Universal indicator (see below) turns purple, blue, yellow, beige, pink or red when it comes into contact with an acid or alkali. The colour shows how weak or strong the acid is.

Strong and weak acids and alkalis

The strength of an acid or alkali does not describe whether the solution is dilute or concentrated. It describes the ability of a substance to form particles called ions. Acids form hydrogen ions and alkalis form hydroxide ions. A strong acid forms a large number of hydrogen ions in solution and a weak acid forms a small number of hydrogen ions in solution. A strong alkali forms a large number of hydroxide ions in solution and a weak alkali forms a small number of hydroxide ions in solution. The strength of an acid or alkali is measured on the pH scale. On this scale the strongest acid is 0 and the strongest alkali is 14. A solution with a pH of 7 is neutral. It is neither an acid nor an alkali. A strong acid has a pH of 0–2, a weak acid has a pH of 3–6; a weak alkali has a pH of 8–11 and a strong alkali has a pH of 12–14.

An electrical instrument called a pH meter is used to measure the pH of an acid or alkali accurately.

7 Here are some measurements of solutions that were made using a pH meter:
A 0, **B** 11, **C** 6, **D** 3, **E** 13, **F** 8.
a) Which of the solutions are:
  i) acids,
  ii) alkalis?
b) If the solutions were tested with universal indicator paper, what colour would the indicator paper be with each one?
c) Fresh milk has a pH of 6. How do you think the pH would change as it became sour? Explain your answer.

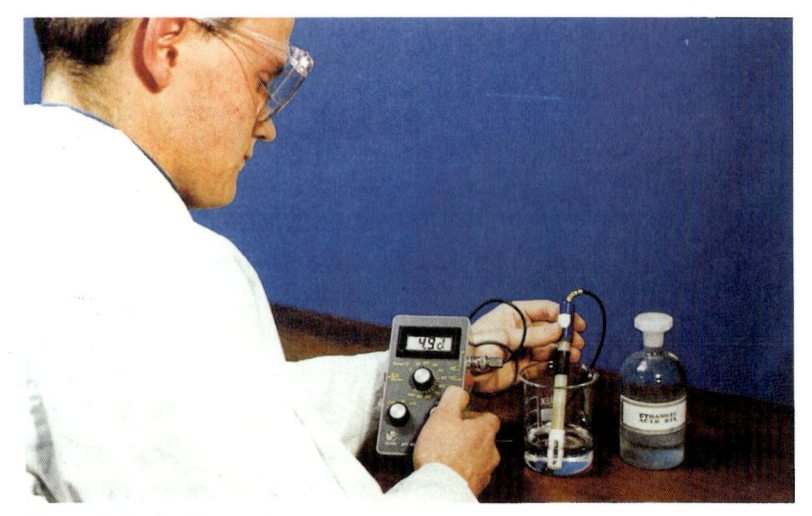

Figure 1.7 A pH meter in use.

For general laboratory use, the pH of an acid or an alkali is measured with universal indicator. This is made from a mixture of indicators. Each indicator changes colour over part of the range of the scale. By combining the indicators, a solution is made that gives various colours over the whole of the pH range (see Figure 1.8).

8 Here are some results of solutions tested with universal indicator paper:
sulphuric acid – red,
metal polish – dark blue,
washing-up liquid – yellow,
milk of magnesia – light blue,
oven cleaner – purple,
car battery acid – pink.
Arrange the solutions in order of their pH, starting with the one with the lowest pH.

9 Identify the strong and weak acids and alkalis from the results shown in questions 7 and 8.

10 Look at page 21 about acids and predict whether nitric acid is a strong or a weak acid. Explain your answer.

11 A sample of acid rain turned universal indicator yellow. What would you expect its pH to be? Is it a strong or a weak acid?

12 Write word equations for the reactions between:
   a) sulphuric acid and zinc oxide,
   b) hydrochloric acid and calcium hydroxide,
   c) nitric acid and calcium carbonate.

13 How is the neutralisation of a carbonate different from the neutralisation of an oxide or a hydroxide?

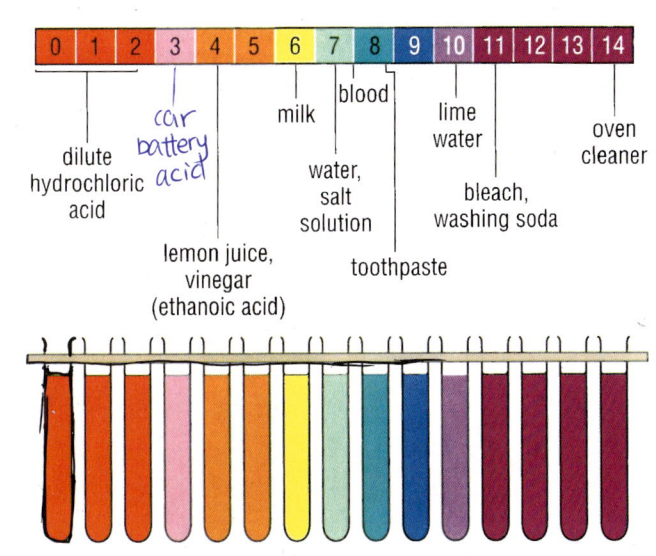

Figure 1.8 The pH scale (top) and universal indicator (bottom).

Neutralisation

When an acid reacts with a base a process called neutralisation occurs in which a salt and water are formed. This reaction can be written as a general word equation:

acid + base → salt + water

Specific examples of neutralisation reactions are:

sulphuric acid + magnesium oxide → magnesium sulphate + water

hydrochloric acid + sodium hydroxide → sodium chloride + water

hydrochloric acid + zinc carbonate → zinc chloride + water + carbon dioxide

nitric acid + sodium hydrogencarbonate → sodium nitrate + water + carbon dioxide

Using neutralisation reactions

When you are stung by a nettle, the burning sensation on your skin is caused by methanoic acid. You can neutralise the acid by rubbing a dock leaf on the skin. As you press the dock leaf against the skin, a base in the leaf juices reacts with the acid in the sting and neutralises it so the burning sensation stops.

A bee sting is acidic and may be neutralised by soap, which is an alkali. A wasp sting is alkaline and may be neutralised with vinegar, which is a weak acid.

ACIDS AND BASES

Sometimes the stomach produces too much acid, which causes indigestion. The acid is neutralised by taking a tablet containing either magnesium hydroxide, calcium carbonate, aluminium hydroxide or sodium hydrogencarbonate (see Figure 1.9).

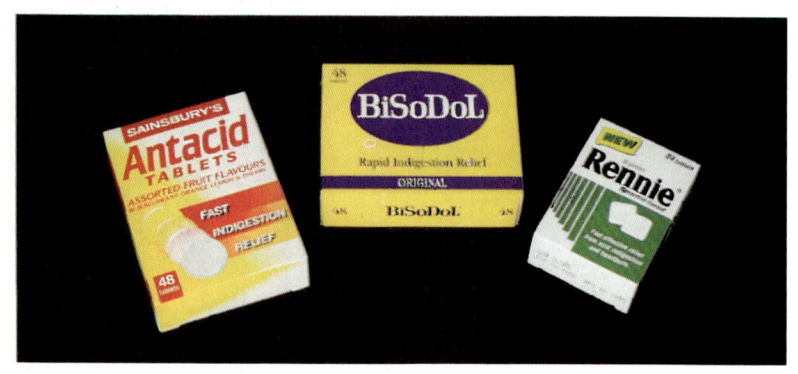

Figure 1.9 A selection of tablets that relieve indigestion.

Acidity in the soil affects the growth of crops. It makes them produce less food. Lime (calcium hydroxide) is used to neutralise acidity in soil. When it is applied to fields it makes them appear temporarily white, as Figure 1.10 shows.

Figure 1.10 Liming fields to improve crop production.

The soda–acid fire extinguisher contains a bottle of sulphuric acid and a solution of sodium hydrogencarbonate (see Figure 1.11). When the plunger is struck or the extinguisher is turned upside down, the acid mixes with the sodium hydrogencarbonate solution and a neutralisation reaction takes place. The pressure of the carbon dioxide produced in the reaction pushes the water out of the extinguisher and onto the fire.

Figure 1.11 A soda–acid fire extinguisher.

ACIDS AND BASES

Acid rain makers

Sulphur dioxide is produced by the combustion of sulphur in a fuel when the fuel is burned. Sulphur dioxide reacts with water vapour and oxygen in the air to form sulphuric acid. This may fall to the ground as acid rain or acid snow.

Oxides of nitrogen are converted to nitric acid in the atmosphere and this falls to the ground as acid rain or acid snow.

The effect of acid rain

When acid rain reaches the ground it drains into the soil, dissolves some of the minerals there and carries them away. This process is called leaching. Some of the minerals are needed for the healthy growth of plants. Without the minerals the plants become stunted and may die (see Figure 1.12).

Figure 1.12 Spruce trees in Bulgaria damaged by acid rain.

The acid rain drains into rivers and lakes and lowers the pH of the water. Many forms of water life are sensitive to the pH of the water and cannot survive if it is too acidic. If the pH changes, they die and the animals that feed on them, such as fish, may also die.

Acid rain leaches aluminium ions out of the soil. If they reach a high concentration in the water the gills of fish are affected. It causes the fish to suffocate.

ACIDS AND BASES

14 In the Arctic regions, snow lies on the ground all winter. As spring approaches and the air warms up, some of the water in the snow evaporates. Later, all the snow melts.

   a) How does the evaporation of the water in the snow affect the concentrations of acids in the snow?

   b) The table below shows how the pH of a river in the Arctic may change during the spring.

Week	pH
1	7.1
2	7.0
3	6.9
4	6.8
5	5.5
6	5.0
7	4.7
8	5.1
9	5.5
10	5.9

   i) Plot a graph of the data.

   ii) Why do you think the pH changed in weeks 5–7?

   iii) Why do you think the pH changed in weeks 8–10?

   iv) How do you expect the pH to change in the next few weeks after week 10? Explain your answer.

Chemical removal of sulphur dioxide

Sulphur dioxide can be removed from the waste gases of factory chimneys in two ways to form useful products.

Lime (calcium hydroxide) can be sprayed into the waste gases where it combines with sulphur dioxide to form calcium sulphate. This rocky material can be used in making the foundation layer of roads.

Ammonia can be mixed with waste gases where it reacts with sulphur dioxide to form ammonium sulphate, which can be used as a fertiliser.

♦ SUMMARY ♦

- Some acids are made by living things (*see page 20*).
- Ethanoic acid in vinegar is made by the oxidation of ethanol in wine (*see page 21*).
- The mineral acids are nitric acid, sulphuric acid and hydrochloric acid (*see page 21*).
- Bases are metal oxides, hydroxides, carbonates and hydrogencarbonates (*see page 22*).
- Bases that dissolve in water are called alkalis (*see page 22*).
- An acid can be detected by its ability to turn blue litmus paper red (*see page 23*).
- An alkali can be detected by its ability to turn red litmus paper blue (*see page 23*).
- The strength of an acid or an alkali depends on the number of ions it contains (*see page 24*).
- The pH scale is used to measure the degree of acidity or alkalinity of a liquid (*see page 24*).
- When an acid reacts with a base a neutralisation reaction takes place (*see page 25*).
- Neutralisation reactions have a wide range of uses (*see page 25*).
- Sulphur dioxide and oxides of nitrogen make acid rain (*see page 27*).
- Useful products can be made from the waste gases of factory chimneys (*see this page*).

End of chapter questions

1. Write an account entitled 'The acids in our lives'.
2. How can you tell when an acid has neutralised an alkali?

2 Physical and chemical changes

Matter everywhere

Figure 2.1 The three states of matter on a school hike.

> **For discussion**
>
> Select one state of matter and imagine that it has been removed from the world. List things that could not exist if it was absent.
>
> Do the same for the other two states of matter.
>
> Would it be possible to live in any of the three imaginary worlds?

When you go for a walk you move across the solid surface of the Earth. Your body pushes through a mixture of gases that we call the air. If it rains as you walk along, droplets of liquid fall from the sky. Solids, liquids and gases are the three states of matter on this planet, and for most other places in the Universe too (see page 81). Not only does your body move through a world made from the three states of matter, it is also made from the three states of matter. Solid bones are moved by solid muscles, while liquids move through your blood vessels and intestines. When you breathe in, air (a mixture of gases) fills your windpipe and lungs.

States of matter

You can tell one state of matter from another by examining its properties.

Solids, liquids and gases all have mass and volume. They also have density, which is found by dividing the mass of the substance by its volume. For example, a solid with a mass of 100 g and a volume of 10 cm^3 has a density of 100/10 = 10 g/cm^3. Another solid with a mass of 200 g and a volume of 10 cm^3 has a density of 200/10 = 20 g/cm^3. This second solid has a higher density than the first solid.

PHYSICAL AND CHEMICAL CHANGES

1 Make a table of the properties of the three states of matter.
2 How are all three states of matter:
   a) similar, and
   b) different?
3 Calculate the densities of these substances:
   a) a plank of wood used for a shelf that has a volume of 1000 cm³ and a mass of 650 g,
   b) the petrol in a car petrol tank that has a volume of 3000 cm³ and a mass of 2400 g,
   c) the air in a box of 1000 cm³ that has a mass of 1.3 g.

A solid has a definite shape and a high density. It is very hard to make it flow or to compress (squash) it. A solid has a definite mass and a volume that does not change. A liquid also has a definite mass and volume. Its density is high and it is hard to compress, but it is easy to make it flow. The shape of the liquid varies and depends on the shape of the container holding it. The shape and volume of a gas vary, and it is easy to make it flow and to compress it. A gas has a definite mass and its density is low.

Figure 2.2 A solid, a liquid and a gas.

Using the properties of matter

The different properties of solids, liquids and gases lead to specific uses.

Solids

As solids have fixed shapes and volumes and are hard to compress, they are used to build structures that range in size from tiny machines to office tower blocks.

Figure 2.3 A tower block under construction in Canary Wharf, London.

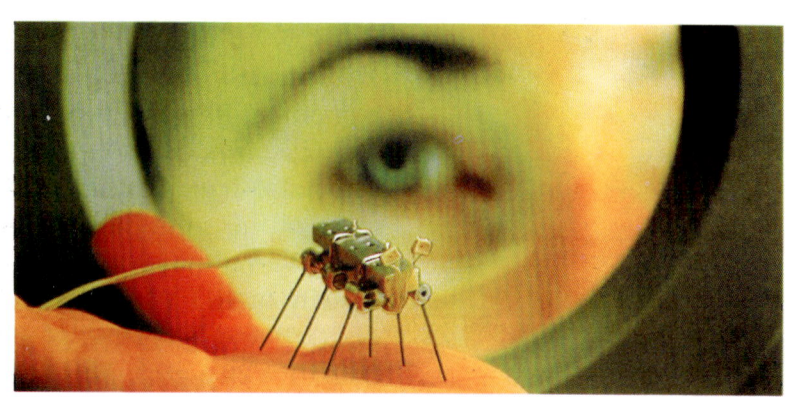

Figure 2.4 A 'robot gnat' developed by nanotechnology.

30

PHYSICAL AND CHEMICAL CHANGES

In structures with moving parts, the solids rub against each other and are worn away. This wearing away is reduced by using a liquid – oil. The oil flows over the surfaces of the moving parts and forms a coating that also moves. This lets the different parts move over each other smoothly without rubbing.

Liquids

A moving car is stopped by pressing the brake pedal with the right foot. Beneath the pedal is a cylinder, which is connected by pipes to four other cylinders – one by the brakes of each wheel. The cylinders and pipes are full of a liquid called brake fluid. When the pedal is pushed down, the force is applied to the liquid in the pipes and cylinders. As the liquid cannot be squashed, it pushes outwards through levers to the brake pads on all four wheels at once. The pads rub against the wheels and slow them down. If the brakes did not work together the car would skid out of control.

Figure 2.5 Oil is used in this engine to prevent wear.

Gases

If a gas is squashed into a small space and is then released, it spreads out rapidly. A compressed gas is used in an aerosol spray to spread droplets of liquid. When the nozzle of the spray is pressed down, some of the gas is released, causing the liquid in the can to form droplets and spread out. The droplets may contain chemicals to kill flies or to give a pleasant smell to a room.

PHYSICAL AND CHEMICAL CHANGES

Figure 2.6 Droplet suspension as an aerosol is sprayed.

Air is a mixture of gases. To make a bicycle more comfortable to ride, air is compressed into the bicycle tyres. Air is pumped into a tyre to give it strength, but as the tyre moves over the small bumps in the road, they push on the flexible tyre walls and squash some of the air even more. This stops the pushing force of the bumps being transferred to the bicycle and stops the cyclist from being shaken about.

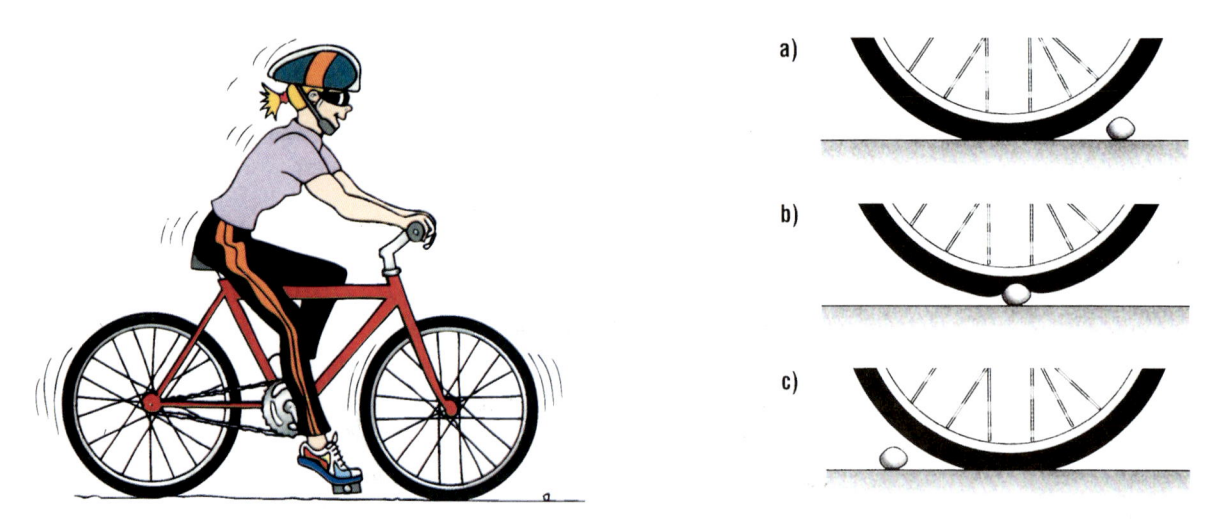

Figure 2.7 A cyclist going over a bump.

PHYSICAL AND CHEMICAL CHANGES

The first ideas about matter

The earliest people used the materials they could find around them such as wood, stone, antlers and skin. When people learned to make fire they began to change one material into another. First they learned how to cook food, then how to bake clay and make pottery and bricks. Eventually they learned how to heat some rocks in charcoal fires so strongly that a chemical reaction took place in which a metal was produced.

By 600 BC, philosophers in the Greek civilisation were thinking about what different things were made of. They were puzzled by the way one substance could be changed into another. They asked the question, 'If a rock can be turned into metal, what really is the rock? Is the rock a kind of metal or is the metal a kind of rock?' They then thought that if one substance could change into another, perhaps it could go on changing into other substances. They did not carry out experiments to test their observations and ideas but tried to explain them with more ideas.

A Greek philosopher called Thales (642–546 BC) believed that all substances were made from different forms of one single substance. He called this substance an element. He observed how water changed from solid to liquid to gas and how plants and animals needed water to stay alive. From these observations he concluded that everything was made from different forms of water.

Other philosophers did not agree with Thales. Some believed that everything was made from air. They believed that air reached up from the ground and filled the whole of space. They thought that air could be squashed to make liquids and solids. Some philosophers suggested that fire was the basic element because it was always changing and it was this element in everything that made things change.

1 Why was the discovery of how to make fire important in making people think about the structure of materials?
2 Why were the Greeks' conclusions about matter not scientific?
3 In what ways do you think Thales saw water change?
4 If a substance was cold and dry, what element did the Greeks think it had?
5 What properties would a material have to show for the Greeks to decide that it contained fire?
6 Which elements do you think the Greeks thought were in:
   a) wood,
   b) oil,
   c) metal?
   Explain your answer.
7 How do you think the Greeks may have explained the changes they saw when a candle burned?
8 Why do you think the Greeks' idea of elements was used for such a long time?

(continued)

33

PHYSICAL AND CHEMICAL CHANGES

Eventually it was agreed that there were four elements from which all matter was made. The elements were water, air, fire and earth. Each element was given properties, and the way that the elements and their properties were related to each other is shown in Figure A.

The Greeks' ideas of the elements were used for 2000 years to explain the structure of materials and the way they change.

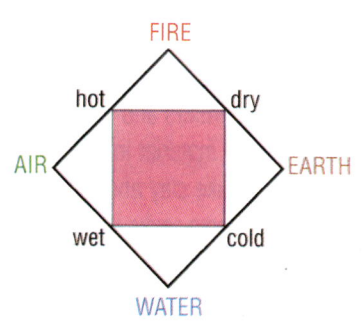

Figure A The Greek elements.

Figure B The four elements – air, water, earth and fire – can be easily identified in our surroundings.

Physical changes

The state of matter of a substance can be changed physically. A physical change of state is a reversible reaction – the reaction can go forwards and also backwards. A physical change of state is brought about by heating or cooling.

Melting and freezing

If a solid is heated enough it loses its shape and starts to flow. This change is called melting and the solid turns into a liquid. The temperature at which melting takes place is called the melting point. This can be found by heating a solid and recording its temperature. When the temperature remains constant or steady the melting point of the solid has been reached.

If a liquid is cooled enough it loses its ability to flow, forms a shape and turns into a solid. This change is called freezing. The temperature at which freezing takes place is called the freezing point. The temperature of the melting point is the same as the temperature of the freezing point.

4 What is the heat source that causes melting of:
 a) ghee,
 b) chocolate in your pocket?
5 Why does the wax not freeze at the top of the candle?
6 Plot a graph for this data.

Time (mins)	Temp. °C
0	0
1	10
2	20
3	30
4	40
5	50
6	55
7	57
8	59
9	60
10	60

Figure 2.8 Icicles with water dripping from them.

PHYSICAL AND CHEMICAL CHANGES

When ice warms in the sunshine of a winter's day in northern Europe, it melts. At night, when the temperature drops below 0 °C, pools of water freeze and turn to ice.

In winter in the polar regions, the temperature remains below 0 °C for many months and large sheets of ice build up over the sea. In the polar summers some of this ice melts.

Figure 2.9 Antarctic sea ice.

For discussion

Many people believe that a change in the climate is taking place now. One consequence of this is that more ice in the polar regions will melt and the sea levels will rise. Some people believe that one of the reasons for the melting of the polar ice is the 'greenhouse effect', where extra carbon dioxide in the air, due to traffic and industry, is trapping more of the Sun's heat in the atmosphere. How will the rise in the sea levels affect people living near sea level close to the coast? How will it affect people living further away?

Melting occurs in many other substances around us. For example, butter melts in a pan during cooking and chocolate can melt in your pocket.

For discussion

What are the current views on climate change? Will it affect the area where you live? Will there be any advantages or disadvantages?

Figure 2.10 Ghee being melted in Indian cookery.

PHYSICAL AND CHEMICAL CHANGES

Figure 2.11 A molten chocolate mess!

While we easily recognise a substance melting, we may not think of a molten substance turning into a solid as freezing. For example, when wax runs down the side of a candle it freezes and becomes a solid before it reaches the bottom.

Figure 2.12 As the wax running down the candles' sides cools, it freezes to become a solid.

Evaporating and boiling

A solid turns into a liquid at one definite temperature but a liquid turns into a gas over a range of temperatures. For example, a drop of water can turn into a gas at room temperature of about 20 °C while outside a puddle of water dries up in the warmth of the Sun. The process by which a liquid changes into a gas over a range of temperatures is called evaporation. The gas escapes from the surface of the liquid. If the temperature of the liquid is raised it evaporates faster.

At a certain temperature the gas forms inside the liquid and makes bubbles which rise to the surface and burst into the air. This process is called boiling. The temperature at which it takes place is called the boiling point. If the boiling liquid is heated more strongly its temperature does not rise but it boils more quickly. The boiling point can be found by heating a liquid and recording its temperature. When the temperature stops rising and remains constant the boiling point of the liquid has been reached (see Figure 2.13).

36

PHYSICAL AND CHEMICAL CHANGES

Reading from the graph you have drawn in answer to question 6, answer the following.

7 What was the temperature of the liquid at:
   a) the start of the experiment,
   b) after 2 minutes?

8 At what time was the temperature:
   a) 30°C,
   b) 70°C?

9 What is the boiling point of the liquid?

10 How did the rate at which the temperature increased change as it reached the boiling point?

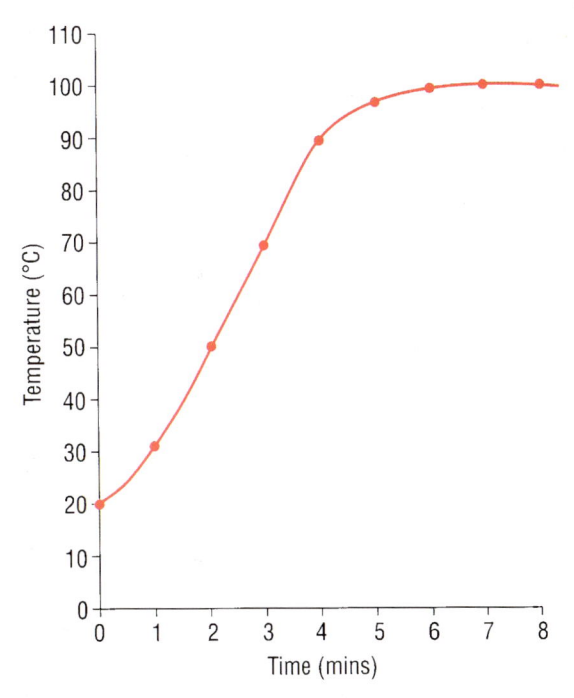

Figure 2.13 Graph to show the boiling point of a liquid.

Condensation

When a gas is cooled down it turns into a liquid by a process called condensation. This process is the opposite of evaporation.

When the water in a kettle boils it forms a colourless gas called steam that rushes out of the kettle spout (see Figure 2.14). A few centimetres above the spout the steam cools and condenses to form a cloud of water droplets which is often wrongly called steam. The real steam cannot be seen and is in the gap between the spout and the base of the cloud of water droplets.

Figure 2.14 A boiling kettle. The steam is just above the kettle spout.

PHYSICAL AND CHEMICAL CHANGES

The changing state of water

It has been estimated that there are 1.5 million million million litres of water on the Earth. Water can change from solid to liquid to gas and back to liquid and solid again at the temperatures found naturally on the Earth. Water moves between the oceans, atmosphere and land in a huge circular path called the water cycle (see Figure 2.15).

Water turns into a gas called water vapour by evaporation at any water surface. In the cool, upper air the water vapour condenses to form millions of water droplets that make the clouds. At the tops of the clouds it is so cold that the droplets freeze and form snowflakes. They fall through the cloud and melt to form raindrops. Falling water in the form of rain, snow or hail is called precipitation. Plant roots take up the water that passes through the soil, and their leaves return water to the atmosphere by transpiration.

11 Water vapour can also condense on the ground at night. What is this condensation called? If this substance freezes we give it another name. What is it?

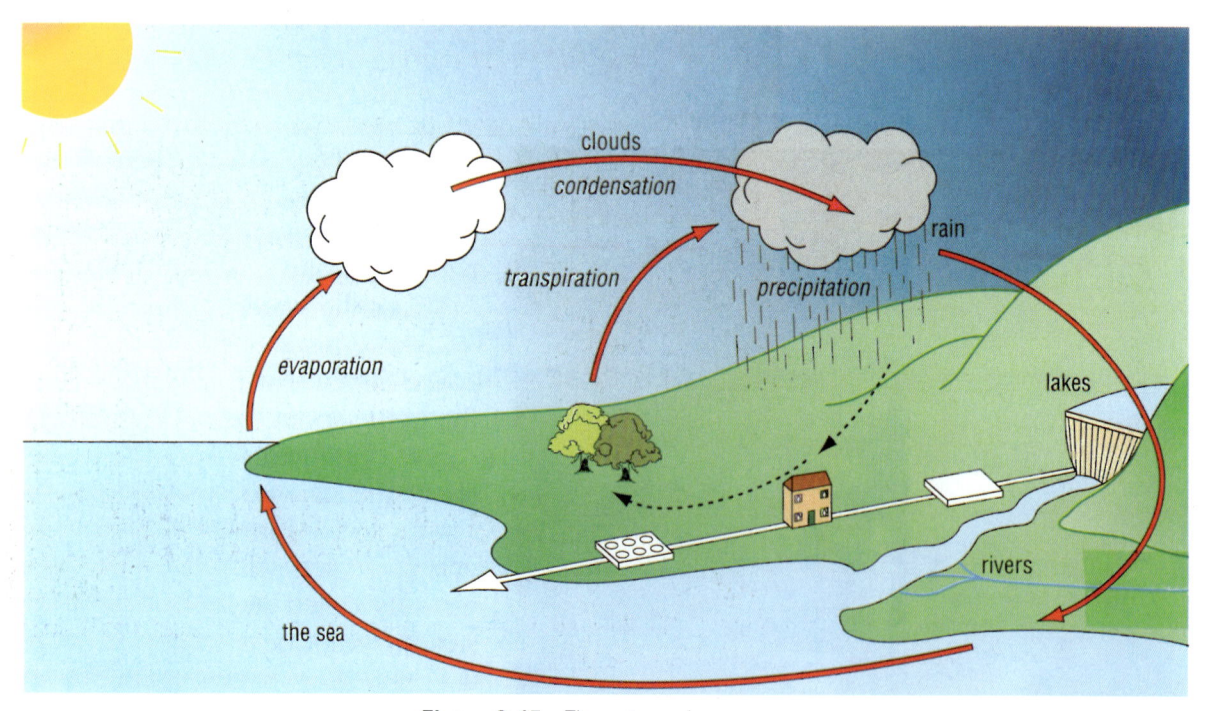

Figure 2.15 The water cycle.

Sublimation

There are a few solids which turn directly into a gas when they are heated. They do not change into a liquid first. This process is called sublimation.

Solid carbon dioxide, known as dry ice, sublimes when it is heated to −78 °C. It can be used on a stage to produce a mist in the air when it warms up (see Figure 2.16).

PHYSICAL AND CHEMICAL CHANGES

Figure 2.16 Dry ice being used at a concert.

In the science laboratory iodine changes from a solid to a gas when heated (see Figure 2.17).

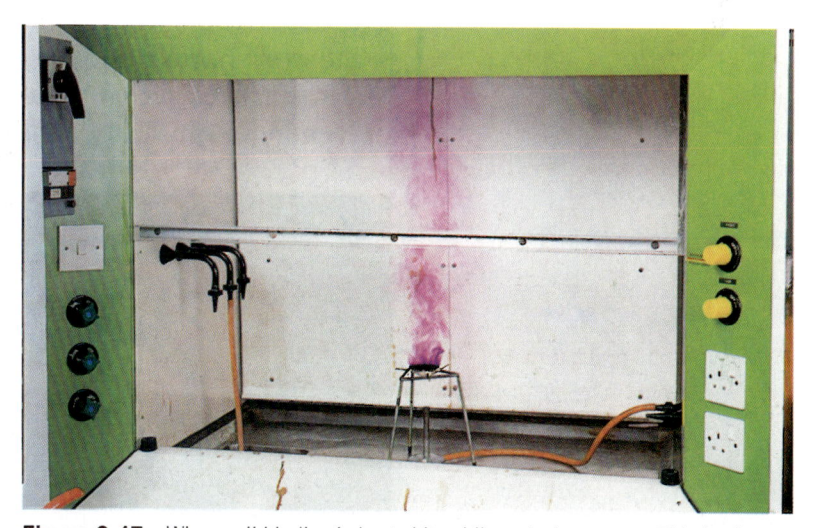

Figure 2.17 When solid iodine is heated it sublimes to form a gas. This is always done in a fume cupboard.

The term sublimation is also used when a gas turns directly into a solid. Sulphur vapour escaping from a volcano sublimes to form a solid crust on the rocks close by (see page 57).

PHYSICAL AND CHEMICAL CHANGES

12 Make a diagram to show the states of matter and the processes that change them. Start by copying out Figure 2.18. Then add the words evaporating, melting, boiling, condensing, subliming and freezing to the appropriate arrows.

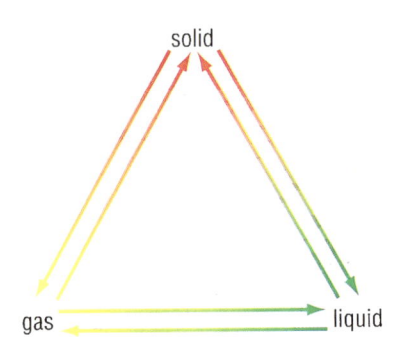

Figure 2.18 The interaction of the states of matter.

Mass and the changes of state

When any substance changes state, such as turning from a liquid to a solid or a liquid to a gas, the mass of the substance does not change.

Chemical changes

Chemical changes are not reversible. If you fry an egg, you will see the white change from a liquid to a solid. However, when you remove the egg from the pan, the white does not change back to a liquid on your plate. The substances present have reacted chemically preventing a physical change from occurring.

Figure 2.19 A chemical change occurs during the frying process.

Combustion

Combustion is a chemical reaction in which energy is given out as heat. If a flame develops, combustion is then called burning. In burning, energy is also given out as light and sound.

PHYSICAL AND CHEMICAL CHANGES

Burning

When a substance burns heat energy is given out. Particles may arise from the substance as smoke and ash and these may be left behind. The reaction cannot be reversed.

Figure 2.20 The ash and smoke cannot be put back together to make the wood.

The Bunsen burner

This piece of apparatus is used to supply heat in a science laboratory. Methane combines with oxygen in the burning process to produce carbon dioxide and water. The parts of a Bunsen burner are shown in Figure 2.21. The air regulator or collar must be turned to fully close the air hole before the burner is lit. The match should be lit and placed to one side of the top of the chimney before the gas tap is switched on.

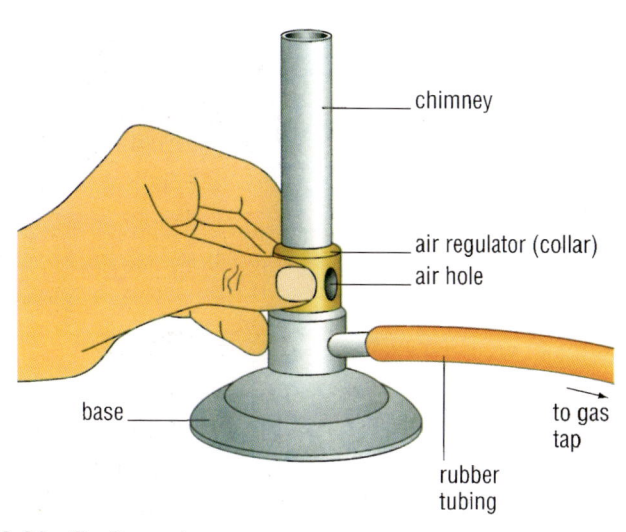

Figure 2.21 The Bunsen burner.

PHYSICAL AND CHEMICAL CHANGES

13 Why is one flame hotter than the other?

14 Why does closing the gas tap a little reduce the size of the flame?

15 What safety precautions should you take when using a Bunsen burner? Explain the reason for each precaution you take.

When the gas is switched on, it shoots out through the small hole, called the jet, and up the chimney. Not all the carbon in the gas combines with the oxygen straight away and carbon particles are produced. They are heated to incandescence and give out a yellow light which makes the flame. If this flame is used to heat anything, the carbon particles form soot on the surface of the apparatus being heated.

The flame produced with the air hole closed is called a luminous flame. It is silent. The carbon in the flame reacts with oxygen in the air and forms carbon dioxide.

If the collar is turned and the air hole is fully opened, air mixes with the gas in the chimney. The gases rush up the chimney and form the blue cone of unburnt gas at the top of the chimney. Above the cone, the complete combustion of methane takes place. The flame made when the air hole is completely open is non-luminous and makes a roaring sound.

Less heat energy is released by the luminous flame than the non-luminous flame because the carbon does not all react with oxygen at once. The hottest part of the non-luminous flame is a few millimetres above the tip of the blue cone of unburnt gas.

The size of the flame is controlled by the gas tap on the bench. If the tap is fully open a large flame is produced. A smaller flame is produced by partially closing the gas tap.

Triangle of fire

The three essentials for a fire are shown in a triangle in Figure 2.22. Remove any side from the triangle of fire and the fire goes out. When fire fighters are trying to put out a fire they may try and remove one or more of the essentials which make up the sides of the triangle. For example, if the fire is near a pile of wood or rubbish that could provide fuel to keep the fire going, the fire fighters will remove it.

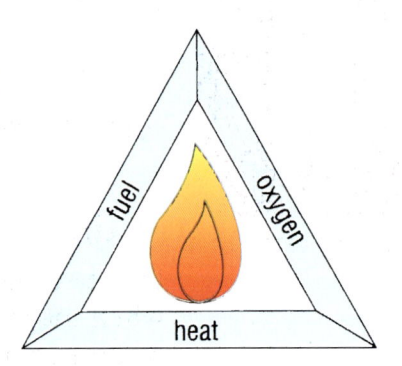

Figure 2.22 The fire triangle.

PHYSICAL AND CHEMICAL CHANGES

Foam is squirted on a fire to form an airtight 'blanket'. This stops oxygen getting to the fire and helps to put it out. Water is used to reduce the amount of heat and to make the fuel too cool to burn. Water and foam should not be used on electrical appliances that are on fire because they can conduct electricity and could give an electric shock to the fire fighters.

Water must not be used on burning oil or petrol because the water sinks below them. When the water boils, the bubbles break through the oil or petrol and spray it over a wide area. Pouring water on the burning oil in a chip pan would cause burning oil to be sprayed out of the pan – this could set fire to the rest of the kitchen. The fire in the oil can be extinguished by covering the top of the pan with a fire blanket.

16 After a road accident, petrol and oil that have spilled onto the road are covered with sand. Why?

Thermal decomposition

When some substances are heated they take part in a chemical reaction called thermal decomposition. An example is heating copper carbonate.

Heating copper carbonate

Copper carbonate is a green powder. If it is heated strongly it breaks down into black copper oxide and carbon dioxide.

The word equation for this reaction is:

copper carbonate → copper oxide + carbon dioxide

Figure 2.23 Before (left) and after (right) the heating of copper carbonate.

Figure 2.24 Magnesium being burnt in a laboratory.

Heating magnesium metal

In the science laboratory a chemical change takes place when magnesium is burnt. The grey metal becomes a white powder (see Figure 2.24).

PHYSICAL AND CHEMICAL CHANGES

♦ SUMMARY ♦

- There are three kinds or states of matter. They are solid, liquid and gas (*see page 29*).
- Each state of matter has properties which are different from other states (*see page 30*).
- The properties of matter have their uses (*see page 30*).
- In reversible physical reactions matter can be changed from one state to another by the processes of melting, freezing, evaporation, boiling, condensation and sublimation (*see page 34*).
- The water cycle is where water undergoes reversible physical changes (*see page 38*).
- In irreversible chemical reactions burning is a type of combustion in which a flame is produced (*see page 40*).
- The Bunsen burner is a device which allows the burning of methane to be controlled to supply heat for experiments (*see page 41*).
- A study of the triangle of fire helps in understanding how fires can be controlled (*see page 42*).
- Copper carbonate breaks down to copper oxide and carbon dioxide when heated (*see page 43*).
- Magnesium metal undergoes a chemical reaction to form a white powder when heated (*see page 43*).

End of chapter questions

1. Imagine you are a water molecule in water vapour in the air above the sea. Explain what might happen to you in the future.
2. When you light a Bunsen burner, what happens to the methane that flows into it?
3. Why do foresters make a wide gap, called a firebreak, between groups of trees in a forest?

3 Investigating everyday materials

Feel the material on your sleeve. Is it hard or soft, rough or smooth, flexible or rigid? When you investigate a material like this you are examining its physical properties. If you were investigating the chemical properties of a material, you would be finding out how the material burnt when heated or how it changed when treated with an acid.

People usually think of the word 'material' when they think about cloth, such as the clothes you are wearing now. But the word 'material' in science is used to mean any substance around you. The wood in a table or desk is a material. So is the water in a tap or the air around you. Humans have been selecting materials from the world around them from the earliest times and using their properties to help them survive. For example, the waxy coating of a large leaf would make a waterproof shelter in a shower of rain.

Figure 3.1 These children in a rainforest are using leaves for shelter.

1 Imagine you were cast away on a desert island (see Figure 3.2). What materials would you select to help you survive? Explain your answer.

Figure 3.2 A desert island seen from the air.

INVESTIGATING EVERYDAY MATERIALS

Materials and their properties

Every material has a number of properties. Here are some properties of five common groups of materials.

- Wood – hard, strong, low density, opaque (does not let light pass through it), can be shaped by carving, rots in water, burns when heated.

Figure 3.3 This canoe has been made by cutting the centre out of a tree trunk. The low density of the wood allows it to float on the water.

- Metal – shiny, hard, strong, malleable (can be shaped by pushing and pulling), ductile (can be pulled into a wire), opaque. Melts when heated very strongly.
- Plastic – melts or burns when heated, low density, can be opaque or translucent (lets some light through but scatters it so that nothing can be seen). Hard, but plastic foam is soft.
- Glass – hard, brittle, transparent (lets light through without scattering it so that objects can be clearly seen through it) or translucent.
- Pottery – hard, brittle, does not melt when heated strongly. Opaque but some kinds of pottery can be translucent.

Figure 3.4 This frame of steel girders provides strong support for the other materials used to make the building.

Figure 3.6 A range of chemicals is added to glass to give it different colours.

Figure 3.5 Plastic shapes. When plastics melt they can be moulded to make shapes for different purposes.

INVESTIGATING EVERYDAY MATERIALS

Figure 3.7 Pottery being made in a simple kiln.

Properties and material shape

Some materials have properties related to their shape. For example, a block of wood is rigid while a strip of wood is flexible (see page 49).

2 Can metals, plastics and pottery be both rigid and flexible? Explain your answer.

Properties and uses of materials

The properties of materials make them suitable for particular uses. For example, wood is hard and strong which means that it is hard wearing. It also has a low density which means it has a low weight for its size. These properties make wood a suitable material for making doors as they receive a great deal of wear and have to be easy to move (open and shut). They also make wood a suitable material for furniture such as chairs, as it gives support to the weight of people without collapsing yet is light enough in weight for the chairs to be moved easily.

3 Why would you not use metal or pottery to make a door? Explain your answer.
4 What are the properties of leather that make it suitable for making shoes?
5 Why is metal not used to make shoes?

Waterproof materials

Waterproof materials are materials which prevent water passing through them. There are two kinds – water-repellent materials and water-resistant materials. A water-repellent material does not have any holes in it through which water may pass. An example of this kind of material is the rubber used to make wellington boots.

A water-resistant material is made of fibres. There are holes between the fibres through which water could pass. However the fibres are coated in silicones which make the water gather up into large droplets which cannot pass through the material. Water-resistant materials are used to make umbrellas and outdoor jackets and trousers.

Figure 3.8 Large drops of water collecting on top of an umbrella.

47

INVESTIGATING EVERYDAY MATERIALS

Figure 3.9 Kitchen cloths are highly absorbent.

6 Devise an investigation to compare the absorbent properties of different brands of paper towels.

Absorbent materials

An absorbent material has holes in its surface through which water can pass and also has spaces inside where the water can collect.

Absorbent cloths and papers are used to wipe up spills in kitchens and laboratories.

Some kinds of rocks, such as sandstone or limestone, are absorbent. The water passes through gaps in the rocky grains. If the temperature falls below 0 °C the water turns to ice. When this change takes place, the volume of the water increases so the ice pushes with great force on the rocky grains and makes the rock break up.

Bricks also take up water. In places where a house is built on damp ground, a layer of waterproof material is laid on the bricks which make the foundations. It is called a damp-proof course and prevents water in the foundation bricks from passing upwards through all the bricks in a wall. Damp walls would make the house cold and cause fungi to grow in any wooden materials joined to the walls.

Figure 3.10 A damp-proof course.

Brittle materials

A brittle substance is one that breaks suddenly if it is hit or bent. A biscuit or a bar of chocolate is a brittle substance. Just try to bend it and it snaps. Glass is a brittle substance. A strong tap can make it shatter into many pieces with sharp edges. As glass has many useful properties, ways have been developed to reduce the danger due to its brittleness. Some glass is toughened by controlling the way it cools. If the glass is struck, it breaks up into many small pieces which do not have

INVESTIGATING EVERYDAY MATERIALS

sharp edges. Laminated glass is made by placing a transparent sheet of plastic between two sheets of glass. When it is struck the plastic holds the broken pieces of glass together.

7 Is the shell of a hen's egg as brittle as the shell of a duck or goose egg? What investigation could you make to find out?

Figure 3.11 The shattered glass in this car windscreen has stayed in place.

Flexible materials

Flexible materials can be bent or squashed but when the pulling or pushing force is removed they spring back to their original shape. A long, thin piece of wood can be flexible. When it is used to make a bow, it is bent out of shape as the arrow is pulled back on the string. When the arrow is released, the wood springs back to its original shape and as it does so it propels the arrow through the air.

Flexible pieces of wood have been used to make bows for thousands of years and are still used today in some parts of the world.

Figure 3.12 A bushman hunting with a bow and arrow.

INVESTIGATING EVERYDAY MATERIALS

Figure 3.13 Hot metal is hammered into shape on an anvil.

Malleable materials

A malleable material is one that can be shaped by hammering or by pressing without the material cracking. Metals are malleable: they can be shaped by hammering. Plasticine or pottery is malleable as it can be shaped by pressing it out.

Transparent materials

You are looking through a transparent material at these words. Air is a transparent material and so is water but we tend to think of transparent materials as being solids such as glass and plastic. A transparent material is one that lets light pass through it without the light rays being scattered. This means that you can see objects clearly through transparent materials. A translucent material also lets light pass through it but the light rays are scattered. This means that you cannot see objects clearly. Translucent glass is used in bathroom windows. Some kinds of fine porcelain are also translucent.

Figure 3.14 Glass can be used to make walls and lets a great deal of light into a building.

Opaque materials

Opaque materials do not let light pass through them. Most materials are opaque but the materials which are used most for their opaque property are the fabrics used to make curtains. They prevent light entering a bedroom on a sunny morning and prevent people outside seeing into homes in the evening. Opaque materials are also useful in making sun shades (see Figure 3.15, opposite).

INVESTIGATING EVERYDAY MATERIALS

Figure 3.15 This sun shade is made from an opaque material.

Heat conductors

Heat can pass through materials by conduction. All materials are made of particles (see Chapter 4) and in some materials the particles pass heat energy along like children passing a parcel at a party (see Figure 3.16, below). Materials which behave in this way are called good conductors of heat. In some materials the particles do not pass heat from one to the next and these materials are known as bad conductors, or insulators.

8 How could you find out which material is the best conductor of heat using a metal spoon, a plastic spoon, a wooden spoon and a piece of aluminium foil, a bowl of hot water and some butter and a knife?

Figure 3.16 As the heat passes along the metal scroll it starts to glow.

Electrical conductors and insulators

A solid can be tested to find out if it conducts electricity by using a circuit like the one shown in Figure 3.17 (overleaf). The solid to be tested is secured between the pair of crocodile clips and the switch is closed. The lamp lights if the solid conducts electricity. By using this circuit, metals and the non-metal carbon, in the form of graphite, are found to conduct electricity. Other non-metals such as sulphur and solid compounds such as sodium chloride do not conduct electricity.

51

INVESTIGATING EVERYDAY MATERIALS

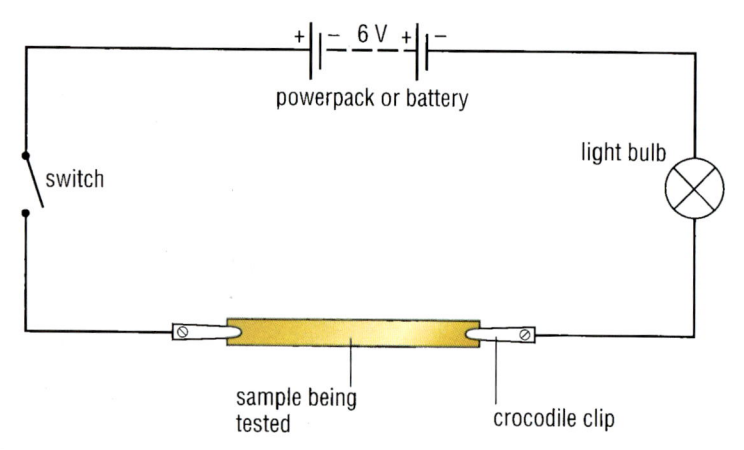

Figure 3.17 A circuit for testing conduction of solid materials.

If a liquid is to be tested, the apparatus shown in Figure 3.18 is used. Graphite (carbon) rods are attached to the crocodile clips. The ends of the rods are then lowered into the liquid. A liquid may be a pure liquid or a solution.

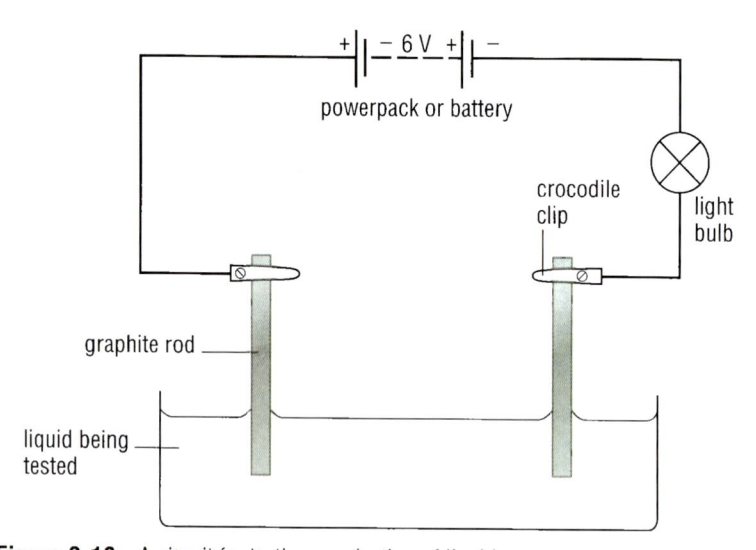

Figure 3.18 A circuit for testing conduction of liquids.

Pure liquids

At room temperature, mercury is a pure liquid that conducts electricity. Ethanol in its pure form does not conduct electricity. Pure liquids which do conduct electricity are formed from compounds such as sodium chloride which have been heated until they melt. Pure water does not conduct electricity but since many substances dissolve readily in it, most water samples are not pure but instead are solutions.

9 It is dangerous to touch electrical switches with wet hands. What may have dissolved in the water from sweaty skin that makes it into a conducting solution?

Solutions

Solutions such as copper sulphate solution and sodium chloride solution, acids such as sulphuric acid and alkalis such as sodium hydroxide solution conduct electricity, but sugar solution does not.

Metals and non-metals

Materials can be divided into two groups – metals and non-metals. The properties of the materials can be used to identify them. Table 3.1 summarises some important physical properties of metallic and non-metallic materials.

Table 3.1 Properties of metallic and non-metallic materials.

Property	Metal	Non-metal
Surface	shiny	dull
Strength	high	high–low
Density	high	high–low
Hardness	high	high–low
Melting point	high	high–low
Conduction of heat	good	poor
Conduction of electricity	good	poor

Note that Table 3.1 refers to metallic materials such as steel and non-metallic materials such as glass and plastic. A further comparison of metallic and non-metallic elements is found in Chapter 9.

There are exceptions to the general properties shown in the table. Graphite, a form of carbon and a non-metal, conducts electricity. Iodine has a shiny surface and looks metallic but it is a non-metal. When it is heated it does not melt but sublimes (see page 39) and turns into a gas. Magnesium, a metal, has a dull surface and when it is heated catches fire and burns with a brilliant white flame (see page 43).

INVESTIGATING EVERYDAY MATERIALS

♦ SUMMARY ♦

- Every material has a number of properties (*see page 46*).
- The shape of a material can affect its properties (*see page 47*).
- Waterproof materials prevent water passing through them (*see page 47*).
- Absorbent materials soak up liquids (*see page 48*).
- Brittle materials break suddenly when hit and bent (*see page 48*).
- Flexible materials can be bent or squashed and then return to their original shape (*see page 49*).
- Malleable materials can be shaped by hammering or pressing (*see page 50*).
- Transparent materials let light pass through them (*see page 50*).
- Opaque materials do not let light pass through them (*see page 50*).
- Heat conductors allow heat to pass through them (*see page 51*).
- Some solids can conduct electricity (*see page 51*).
- Some liquids can conduct electricity (*see page 52*).
- Materials can be divided into two groups – metals and non-metals (*see page 53*).

End of chapter questions

1. Here are some items you may find in a kitchen: a metal saucepan, a wooden spoon, a glass measuring jug and a cotton cloth.
   Here are some properties of materials: absorbent, a good conductor of heat, a good heat insulator, transparent.
   Match each property with a material and say how the property of sx material helps you make a meal.
2. How many different materials is your shoe made from? What are the properties of each material? How are these properties useful?

4 Particle theory

Particles of matter

Observations on the three states of matter and how they change can be explained by considering that matter is made of particles. This is called the 'particle theory of matter'.

Particles in the three states of matter

In solids, strong forces hold the particles together in a three-dimensional structure. In many solids the particles form an orderly arrangement called a lattice. The particles in all solids move a little. They do not change position but vibrate to and fro about one position.

In liquids, the forces that hold the particles together are weaker than in solids. The particles in a liquid can change position by sliding over each other.

In gases, the forces of attraction between the particles are very small and the particles can move away from each other and travel in all directions. When they hit each other or the surface of their container they bounce and change direction.

1 According to the particle theory, why do liquids flow but solids do not?
2 How is the movement of particles in gases different from the movement of particles in liquids?

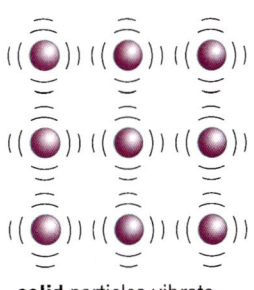

solid particles vibrate to and fro

liquid particles have some freedom and can move around each other

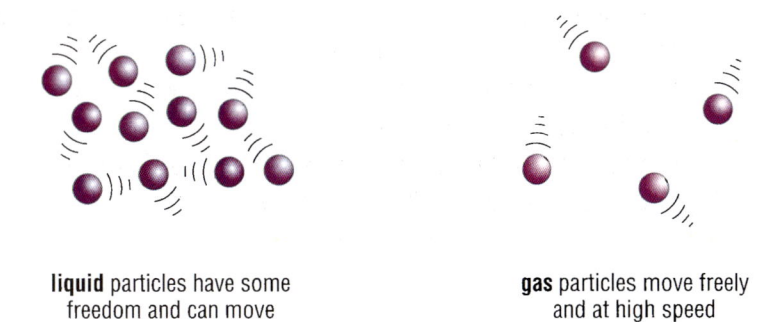

gas particles move freely and at high speed

Figure 4.1 Arrangement of particles in a solid, a liquid and a gas.

When matter changes state

Expanding and melting

If a solid is heated, it expands and then melts. The heat provides the particles with more energy. The energy makes the particles vibrate more strongly and push each other a little further apart – the solid expands. If the solid is heated further, the energy makes the particles vibrate so strongly that they slide over each other and become a liquid. During the time from when the solid starts to melt

PARTICLE THEORY

until it has completely turned into a liquid its temperature does not rise. All the heat energy is used to separate the particles so that they can flow over one another.

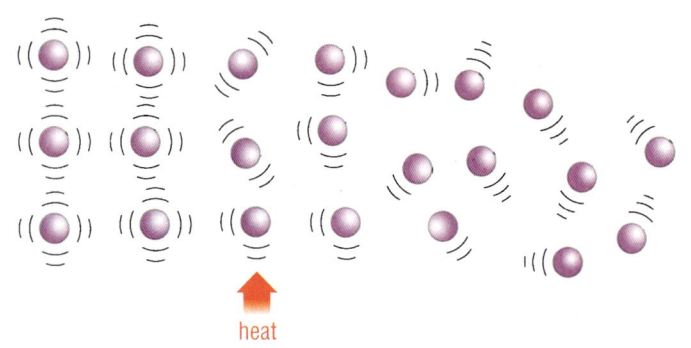

Figure 4.2 The particle arrangement in a solid (on left) changes as the heat turns it into a liquid (on right).

Freezing

If a liquid is cooled sufficiently the particles lose so much energy that they can no longer slide over each other. The only movement possible is the vibration to and fro about one position in the lattice. The liquid has become a solid.

Figure 4.3 The water in this waterfall has frozen to form ice.

Evaporation

The particles in a liquid have different amounts of energy. The particles with the most energy move the fastest. High energy liquid particles near the surface move so fast that they can break through the surface and escape into the air and form a gas.

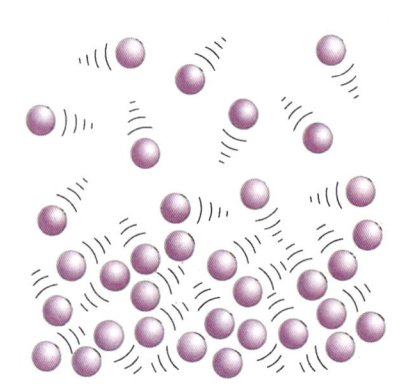

Figure 4.4 Evaporation.

56

PARTICLE THEORY

Boiling

When a liquid is heated all the particles receive more energy and move more quickly. The fastest moving particles escape from the liquid surface or collect in the liquid to form bubbles. The bubbles rise to the surface and burst open into the air. The fast moving particles released from the liquid form a gas.

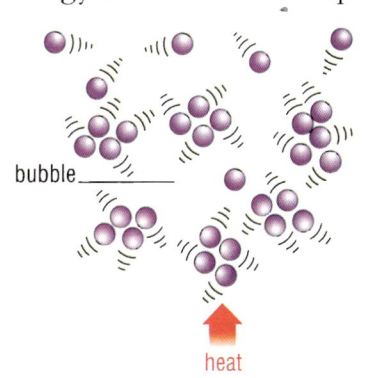

Figure 4.5 Boiling.

Condensation

The particles in a gas possess a large amount of energy which they use to move. If the particles are cooled they lose some of their energy and slow down. If the gas is cooled sufficiently, the particles lose so much energy that they can no longer bounce off each other when they meet. The particles now slide over each other and form a liquid.

Sublimation

When a few substances, such as solid carbon dioxide (see pages 38–39) and iodine (see page 39) are heated, the energy the particles receive makes them separate and form a gas without forming a liquid first. This is called sublimation. Sublimation also occurs when a gas turns into a solid (see Figure 4.7).

Figure 4.6 Breathing onto a cold window causes water vapour in your breath to condense.

3 How is melting different from evaporation?
4 How is boiling different from sublimation?
5 How are condensation and freezing similar?

Figure 4.7 The sulphur gas has sublimed to a solid as it cools after leaving the volcano.

57

PARTICLE THEORY

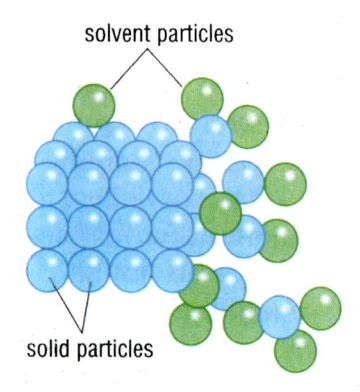

solvent particles

solid particles

Figure 4.8 Dissolving.

Dissolving

There are small gaps between the particles in a liquid. When a substance dissolves in a liquid, its particles spread out and fill the gaps. Figure 4.8 shows how particles in a solid are pulled apart by the particles in the liquid (called the solvent) and move between the particles of the liquid. When a substance dissolves in a liquid, it forms a solution (see page 64).

Pressure

Solids can generate pressure – think of a brick pressing down on your toes. Liquids can generate pressure too. A dam has to be built with thick walls to withstand the pressure of the water that collects in the reservoir behind it.

A gas does not have a surface like a solid or a liquid but it still pushes on any surface with which it makes contact. This push on the surface area of a liquid or solid is called pressure.

A gas contains millions of quickly moving particles. Every second, large numbers are bouncing off the walls of the gas container. The force of these particles as they push against the surface gives rise to the gas pressure.

If the gas is heated the particles move faster and bounce off the container surface more frequently and with more force, so the gas pressure rises. When the gas is cooled the particles move more slowly. They bounce off the container's surface less frequently and with less force, and the gas pressure falls.

When a gas is squashed into a smaller volume but its temperature is kept the same, as shown in Figure 4.9, the particles have less space in which to move. They bounce off the container walls more frequently and the gas pressure rises.

6 What two things can make the pressure of a gas rise?

7 a) What happens to the gas pressure if the gas is released from a small container into a large container?

b) Why does the gas pressure in **a)** change?

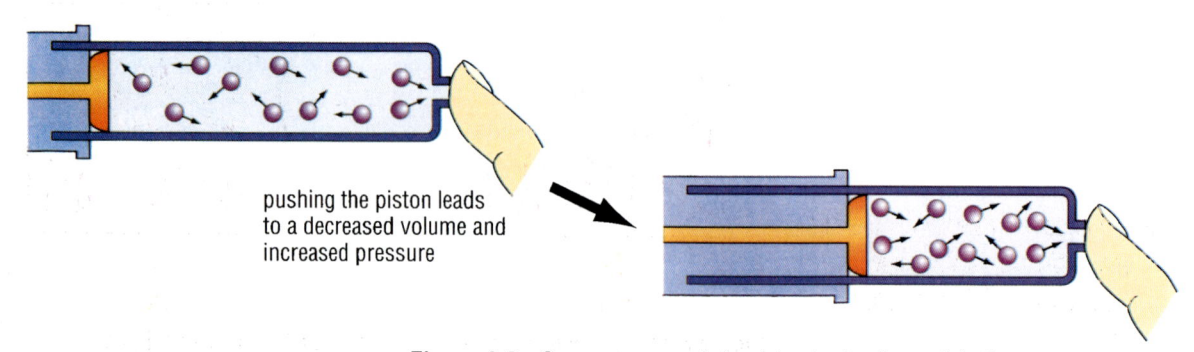

pushing the piston leads to a decreased volume and increased pressure

Figure 4.9 Gas pressure can be explained using the particle theory.

PARTICLE THEORY

Pressure and changes of state

The state of matter of a substance can be changed by changing the pressure acting on it. Under very high pressure a gas can be turned into a liquid or a liquid turned into a solid.

Pressure on ice

As skaters move across the ice their weight pushes down through the small surface of the blades and makes a large pressure. The ice beneath the blade melts. When the skaters have passed by, the pressure is reduced on the ice surface and the water freezes again. This change happens because ice is less dense than its liquid form, water. The change does not happen with other solids because they are denser than the liquids they form when they melt.

Atmospheric pressure

The atmosphere is a mixture of gases that covers the surface of the Earth. The atmosphere is 1000 km thick and pushes on every square centimetre of the Earth's surface. The pressure of the atmosphere at sea level is called standard pressure and is about 10 N/cm^2. It is the pressure at which the boiling point of any substance is measured. At the top of very high mountains the pressure of the atmosphere is less than at sea level.

Boiling and low pressure

If a flask containing a liquid is connected to a vacuum pump and some of the air above the liquid is sucked out, there is less air inside the flask to push on the surfaces and the air pressure is smaller.

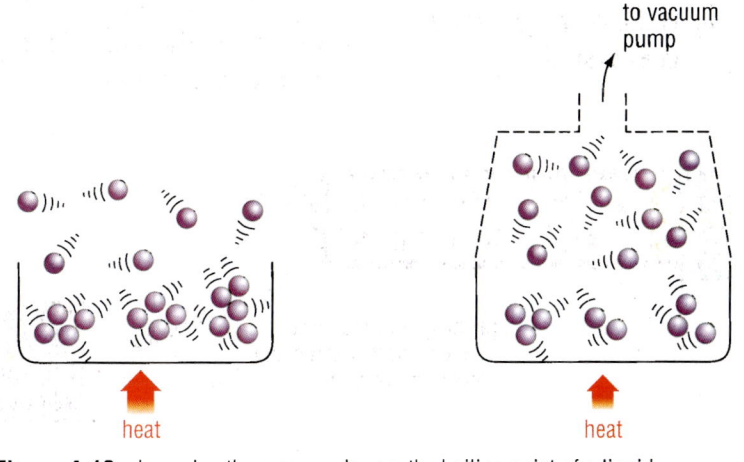

8 If you boiled water at the top of a mountain, would you expect it to boil at 100 °C? Explain your answer.

Figure 4.10 Lowering the pressure lowers the boiling point of a liquid.

PARTICLE THEORY

The reduced air pressure allows evaporation to take place more quickly and less heat is needed to make the liquid boil. Lowering the atmospheric pressure on a liquid lowers its boiling point.

Boiling and high pressure

When a gas gets hot it expands and increases its pressure on the surfaces around it. If water is boiled in a pan with a lid, the steam escaping from the water pushes on the lid and makes it rise – allowing the gas to escape.

Figure 4.11 The lid on this pan of boiling water is being pushed up by the steam.

Diffusion

Diffusion is a process in which one substance spreads out through another. It occurs in liquids and gases. For example, if you put a drop of ink in a beaker of water the ink spreads out through the water by diffusion and colours it (see Figure 4.12). The gases escaping from food cooking in the kitchen can move by diffusion to other rooms in the home. The moving particles in the different liquids flow over each other and the particles in the different gases bounce off each other. These movements eventually spread all the particles of one substance evenly through the other. Liquids are denser than gases and this makes diffusion in liquids much slower than diffusion in gases.

9 Draw diagrams similar to those on pages 56 and 57 to show the following processes:
   a) sublimation,
   b) condensation,
   c) diffusion.

PARTICLE THEORY

At start — After an hour — After a day

Figure 4.12 Black ink diffusing through a beaker of water.

Testing for purity

If water contains other substances dissolved in it, the water is impure. Impure water forms ice that melts at a temperature below 0 °C and boils at a temperature above 100 °C. The melting and boiling points of a substance can therefore also be used to find out if a substance in the laboratory is pure.

♦ SUMMARY ♦

- The particle theory of matter can be used to explain how matter behaves (*see page 55*).
- The particles in the three states of matter behave differently (*see page 55*).
- When the activity of the particles changes, matter changes from one state to another (*see pages 55–57*).
- Changes in gas pressure can be explained by the way the particles push on the sides of their container (*see page 58*).
- The state of matter can be changed by changing the pressure acting on the substance (*see page 59*).
- Diffusion is a process in which one substance spreads out through another (*see page 60*).
- The melting and boiling points of a substance can be used to test for the purity of the substance (*see this page*).

End of chapter questions

1. Use the particle theory of matter to explain what happens to the particles when an ice cube melts and the water it produces evaporates.
2. Why does a bicycle tyre get harder when you pump it up?

5 Mixtures and separating techniques

A mixture is composed of two or more separate substances. The composition of a mixture may vary widely. One mixture of two substances, A and B, might have a large amount of A and a small amount of B. Another mixture might have a small amount of A and a large amount of B.

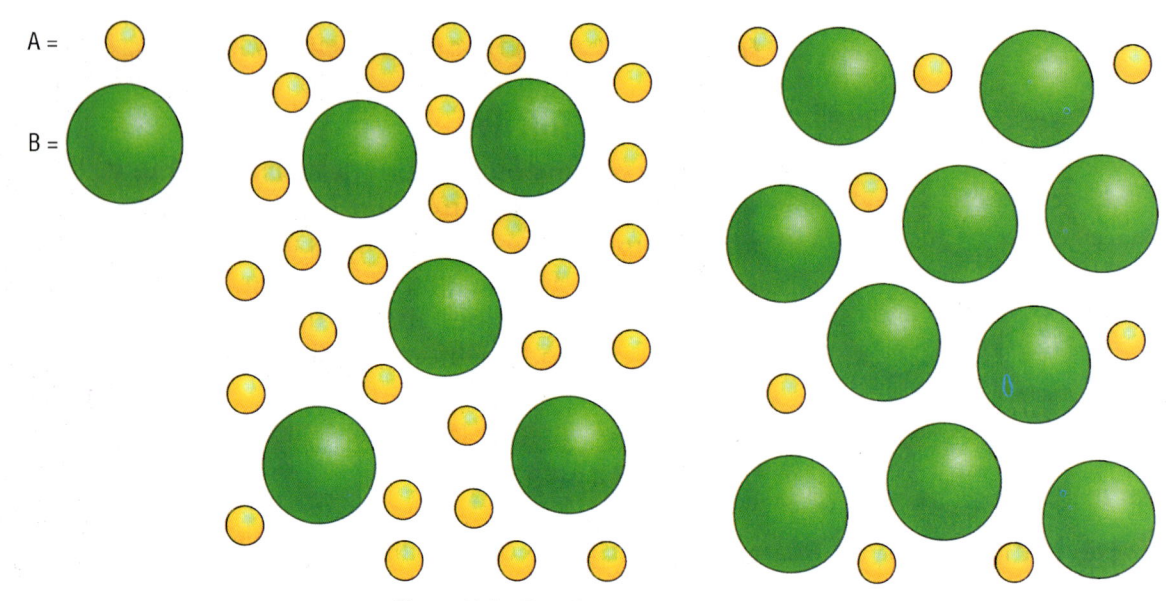

Figure 5.1 Two different mixtures of A and B.

The substances in a mixture can be separated by physically removing one substance from another, as shown in the separating techniques in this chapter.

Different kinds of mixtures

Solid/solid mixtures

Soil is a mixture of different solid particles. Some particles such as clay are very small while others such as sand are larger.

Solid/liquid mixtures

If clay is stirred with water it forms a cloudy mixture. The tiny clay particles are suspended in the water. The mixture is called a suspension. If a solid dissolves in a liquid a solution is made (see page 64).

Figure 5.2 A cross-section of soil.

Solid/gas mixtures

The smoke rising in the hot air from a bonfire contains particles of soot and ash. A mixture of solid and gas also occurs in the dust produced when the wall of a building is being cleaned by blasting a jet of sand at it.

Liquid/liquid mixtures

Milk is a mixture of tiny droplets of fatty oil in water. This kind of mixture is called an emulsion. An emulsion mixture is also found in some kinds of paint.

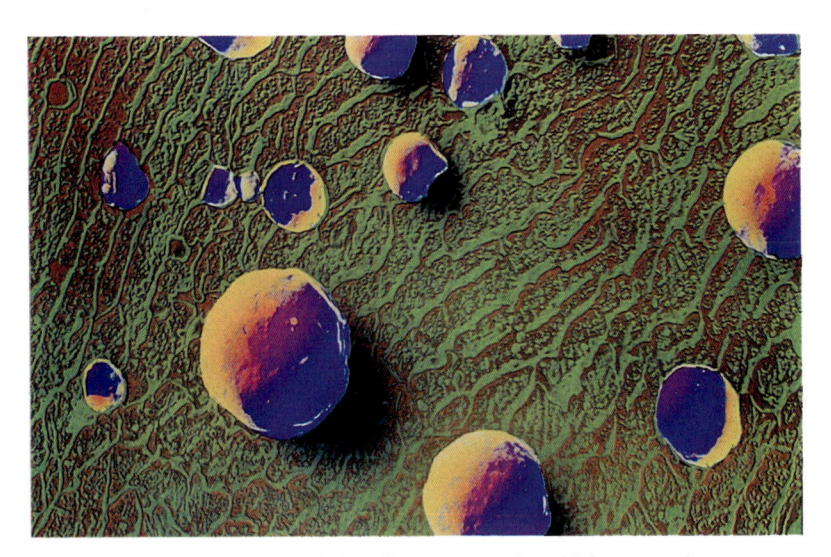

Figure 5.3 Fat globules in whole milk, as seen under a high power microscope.

Gas/gas mixtures

Gases can move freely and when two different gases meet they mix. The most common mixture of gases is the one around you right now – air.

Liquid/gas mixtures

When water vapour in the air condenses above the cool surface of a lake or a field, the tiny droplets form a mist or fog.

When you press the top of an aerosol can, a mist of liquid droplets in a gas (an 'aerosol') is sprayed into the air (see Figure 2.6 page 32).

Gas/liquid mixtures

When bubbles of gas are trapped in a liquid they form a foam. Foam is made when the nozzle of a shaving foam cylinder is pressed. Some products that protect you from sunburn are foams.

MIXTURES AND SEPARATING TECHNIQUES

Solutions

The most common form of a mixture in chemical experiments is the solution. A solution is made when a substance, called a solute, mixes with a liquid, called a solvent, in such a way that the solute can no longer be seen. This type of mixing is called dissolving.

Although the solute cannot be seen it has not taken part in a chemical reaction and can be recovered from the solution by separating it from the solvent. The solute may be a solid, liquid or gas.

A liquid that dissolves in a solvent, water, for example, is said to be miscible with water. A liquid that does not dissolve in a solvent is said to be immiscible with it.

A gas or a solid that dissolves in a solvent is said to be soluble in that solvent. A solid or gas that does not dissolve in a solvent is said to be insoluble in that solvent (see Figure 5.4).

1 What is the difference between a solvent and a solute?
2 What is the difference between a substance that is soluble in water and one that is insoluble in water?
3 What is the difference between an immiscible substance and an emulsion?

Copper sulphate dissolves in water to form a blue solution.

Clay does not dissolve in water, but forms a suspension that settles to the bottom after some time.

Figure 5.4 Soluble and insoluble substances.

MIXTURES AND SEPARATING TECHNIQUES

Saturated solutions

If the temperature of a solvent is kept steady or constant, and the solute is added in small amounts, there comes a time when no more solute will dissolve. The solution is then said to be saturated. If the temperature of the saturated solution is raised, it is able to take in more solute until it becomes saturated at the new temperature.

Solubility

The solubility of a solute in a solvent at a particular temperature is the maximum mass of the solute that will dissolve in 100 g of the solvent, before the solution becomes saturated.

If the temperature of the solvent is raised the solubility of the solute usually increases (see Figure 5.5). If the solubilities of a substance at different temperatures of the solvent are plotted on a graph, a solubility curve is made.

4 What do the solubility curves of the three substances in Figure 5.5 show?

5 How does the solubility of potassium nitrate change when the temperature of the solvent is raised from 30 to 50 °C?

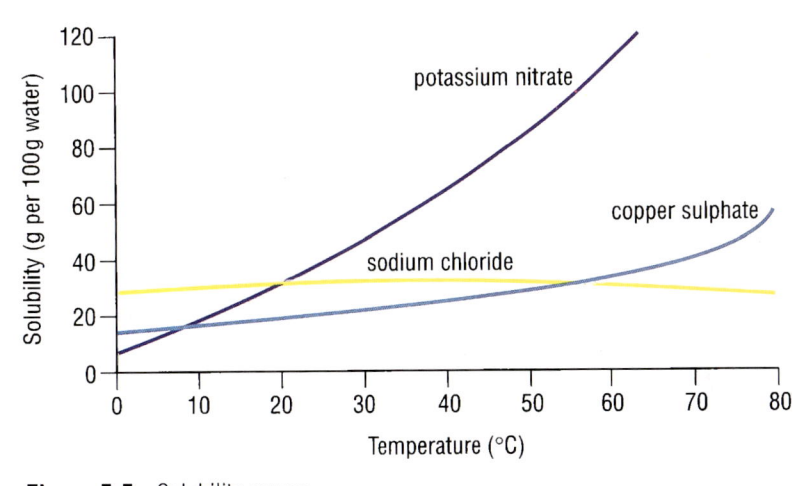

Figure 5.5 Solubility curves.

Liquids and gases in solvents

The miscibility of a liquid with a solvent may change with a change in temperature of the solvent. For example, ethanol and cyclohexane, which are used to make paint remover, form two separate layers when they are cold but become miscible when they are hot.

MIXTURES AND SEPARATING TECHNIQUES

6 How does the solubility of oxygen in water vary with the temperature of the water?

More gas will dissolve in cold water than in warm water – this is the opposite of what happens with solids. Hot water entering a river from a power station can warm the river water so much that not enough oxygen can dissolve in it for the fish to breathe. Some species of water animals can only live where there is a high concentration of oxygen in the water. These species must live in the cool waters of mountain streams.

Different solvents

Water has been called the universal solvent because so many different substances dissolve in it. However, there are many liquids used as solvents in a wide range of products. Ethanol is used in perfumes, aftershaves and glues. Propanone is used to remove nail varnish and grease. Gloss paint is dissolved in white spirit.

Substances that dissolve in one solvent do not necessarily dissolve in others. Salt dissolves in water but not in ethanol. White sugar dissolves in both.

Separating mixtures

The substances in a mixture have not taken part in a chemical reaction and have kept their original characteristics. These characteristics are used in the following techniques to separate the substances.

Separating a solid/solid mixture

A mixture of two solids with particles of different sizes may be separated by using a sieve. The particles in soil are analysed by putting the soil in the top compartment of a soil sieve and shaking it. Each part of the soil sieve has a mesh with smaller holes than the one before. Different sized particles are caught in each layer as the soil moves from the top to the bottom.

Magnetic materials can be separated from non-magnetic materials by passing the materials close to a magnet. In a magnetic separator, the cylinder is a magnet that attracts iron and steel items to it, while other metals fall away. The magnetic materials are then knocked off the cylinder into another collection bay, as shown in Figure 5.7.

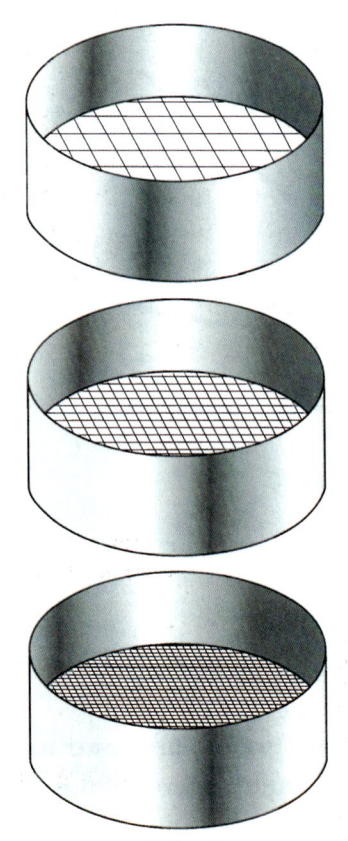

Figure 5.6 A soil sieve.

MIXTURES AND SEPARATING TECHNIQUES

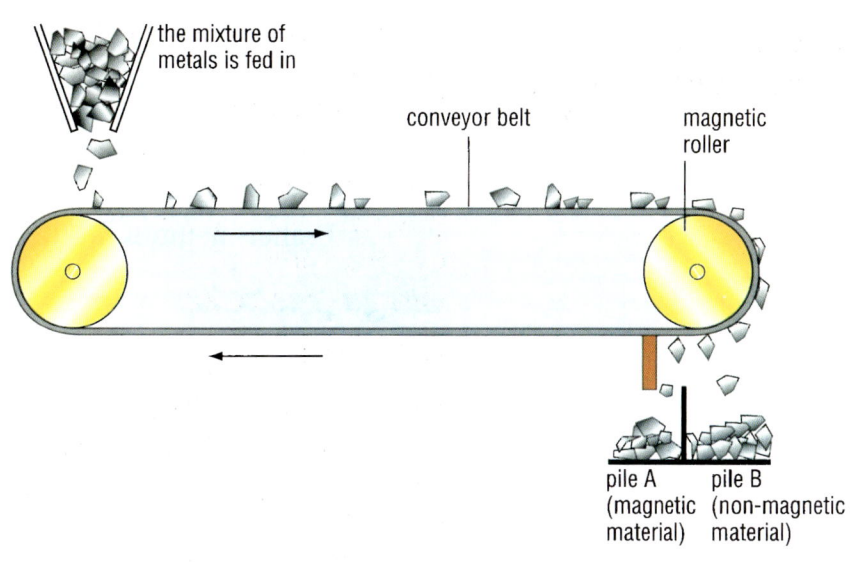

Figure 5.7 A magnetic separator.

Many metals are found in rocks combined with other substances and with a great deal of worthless material. This mixture is known as an ore (see page 110).

Metal compounds and the worthless material can be separated using a flotation cell. The ore is broken up into fragments and added to a mixture of water, oil and a range of chemicals. When compressed air is blown through the chemical mixture, a froth is produced which rises to the surface. The chemicals also help the particles containing the metal to cling to the froth, while the worthless material is left behind. The skimmers at the surface remove the froth and valuable metal compound.

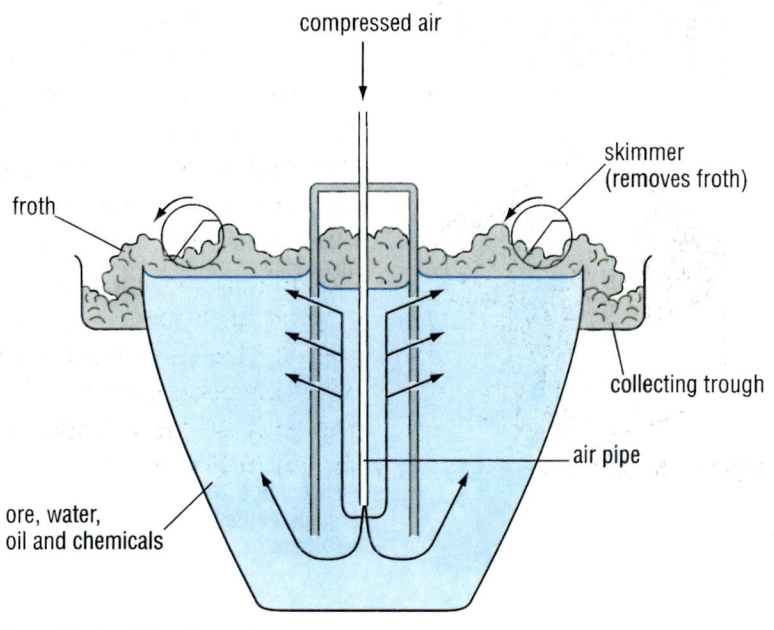

7 What is used in a flotation cell to separate a metal compound from its ore?

Figure 5.8 A flotation cell.

MIXTURES AND SEPARATING TECHNIQUES

Separating an insoluble solid/liquid mixture

In the home, sieves are used to separate insoluble solids, such as peas, from liquids. This is possible because the particle size of the solid is very much larger than that of water. In chemistry, this is not usually the case and other methods are needed.

Large particles

Decanting

Large particles of an insoluble solid in a liquid settle at the bottom of the liquid's container. They form a layer called a sediment. The liquid and solid can be separated by decanting. A liquid is decanted by carefully pouring it out of the container without disturbing the sediment at the bottom. At home, some medicines and aftersun lotions form a sediment in the bottom of the bottle and have to be shaken to mix the solid and liquid before being used.

Figure 5.9 Decanting a liquid from a jug.

Small particles

Filtration

In many laboratory experiments, filtration is carried out by folding a piece of filter paper to make a cone and inserting it in a filter funnel. The funnel is then supported above a collecting vessel and the mixture to be separated is poured into the funnel.

The filter paper is made of a mesh of fibres. It works like a sieve but the holes between the fibres are so small that only liquid can pass through them. The solid particles are left behind on the paper fibres.

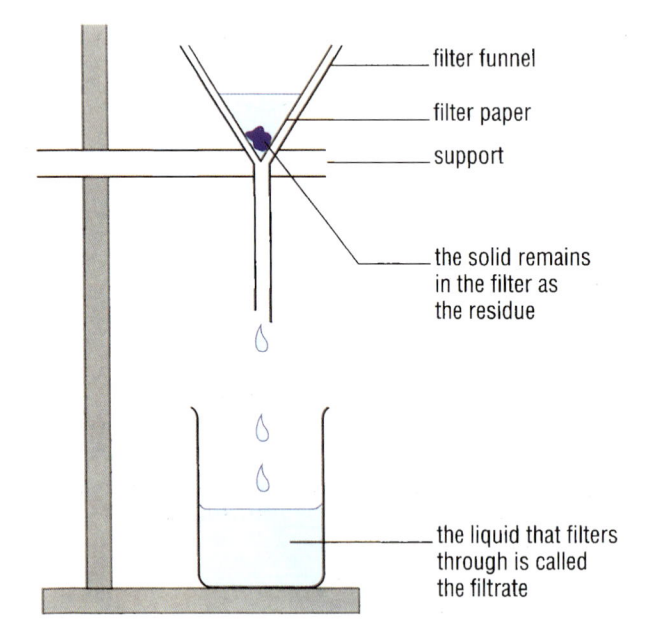

Figure 5.10 Filtration with a filter funnel.

MIXTURES AND SEPARATING TECHNIQUES

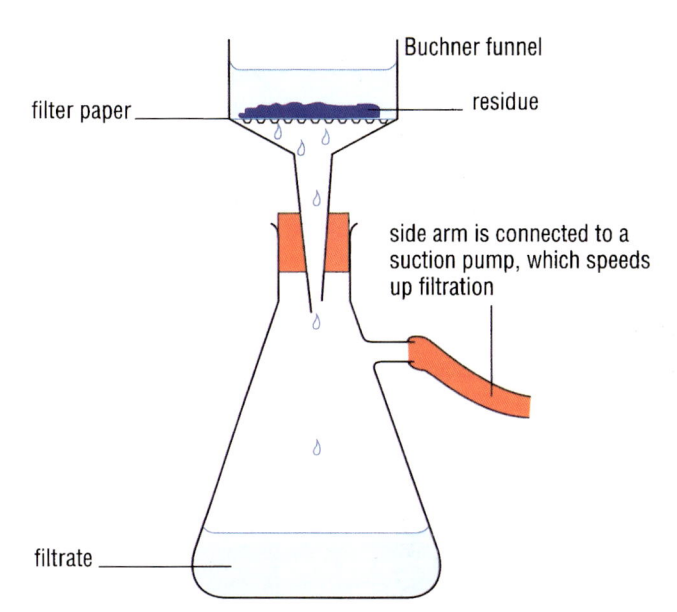

Figure 5.11 Filtration with a Buchner funnel.

A fast filter

A Buchner funnel has holes in it. Filter paper is spread out over the holes. The funnel is fastened into the top of a flask which is connected to a suction pump by a rubber tube. The suction pump draws air out of the flask. When the mixture is poured into the funnel and the suction pump is switched on the air pressure inside the flask is reduced. The higher air pressure above the mixture pushes on it and speeds up filtration.

In both kinds of filtration the substance left behind on the filter paper is called the residue and the liquid that has passed through the filter paper is called the filtrate.

Centrifuge

Very small insoluble particles in a liquid may be separated from it using a centrifuge. This machine has an electric motor which spins several test-tubes mounted on a central shaft. All the test-tubes are carefully balanced by having the same amount of liquid poured into them. As the test-tubes spin, the small particles are forced to the bottom of the test-tubes and form a layer like a sediment. When the test-tubes are removed from the centrifuge the liquid can be decanted from them.

8 What method of separation would you use to separate:
   a) water and fine sand,
   b) water and gravel,
   c) water with very tiny particles floating in it?
   In each case explain how the method of separation works.

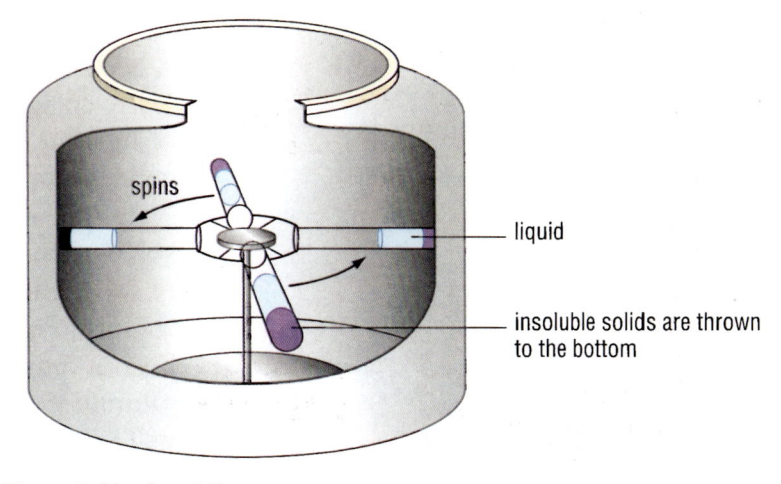

Figure 5.12 A centrifuge.

69

MIXTURES AND SEPARATING TECHNIQUES

Separating a solute from a solute/solvent mixture

These methods can be used to separate a solid solute from a solvent.

Evaporation

If a solution is heated gently, the solvent evaporates from the surface until only the solid is left. Distilled water is made by boiling and condensing water (see page 72) to remove impurities.

Figure 5.13 Copper sulphate solution being evaporated over a water bath.

Tap water and sea water may be compared with distilled water by setting up samples of all three kinds of water and heating them gently until all the liquid has evaporated, leaving only the solid content behind.

Crystallisation

A crystal is a solid structure with flat sides. Many substances form crystals. One way of making crystals is to start with a concentrated solution of a substance.

As the solution is gently heated the solvent evaporates and the concentration of the solute in the solution rises until the solution is saturated (see page 65). If the heat is removed at this time and the saturated solution is left to cool, the solid will form crystals.

Figure 5.14 Crystals of copper sulphate formed in an evaporating dish.

70

MIXTURES AND SEPARATING TECHNIQUES

Separating several different solutes from a solvent – chromatography

A simple chromatography experiment can be performed with filter paper, a dropper, ink and water. A drop of ink is placed in the centre of the filter paper, then a drop of water is placed on top of it. The water dissolves the coloured pigments and spreads out through the filter paper, carrying the pigments with it. Each kind of pigment moves at a different speed to the others so that they spread out into different regions of the paper. They do this because the pigments vary in their solubility and their tendency to stick to the paper.

9 The unknown ink mixture used in Figure 5.15 is suspected of containing substances A, B and C. Do the results confirm this?

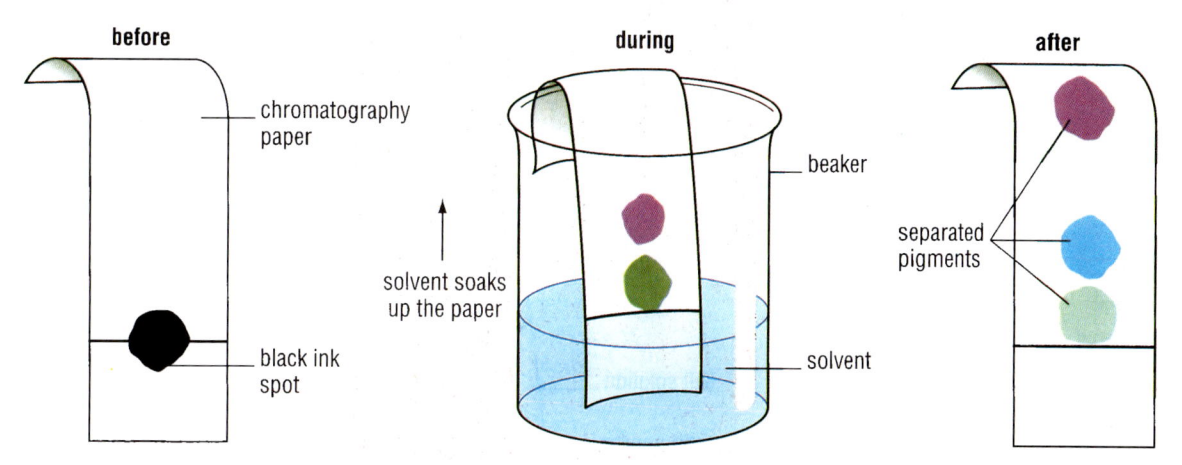

Figure 5.15 Simple paper chromatography.

Very soluble pigments which do not tend to stick to the paper move the furthest and pigments which are not very soluble and tend to stick to the paper move the least. When the separation is complete, the paper is dried. The paper with its separate pigments is called a chromatogram.

Substances that do not dissolve in water can be separated by chromatography by using other solvents, such as propanone. When one of these solvents is used the chromatography paper is enclosed in a tank. This makes sure that the solvent vapour does not escape but surrounds the paper keeping it saturated with solvent and helping the substances to separate.

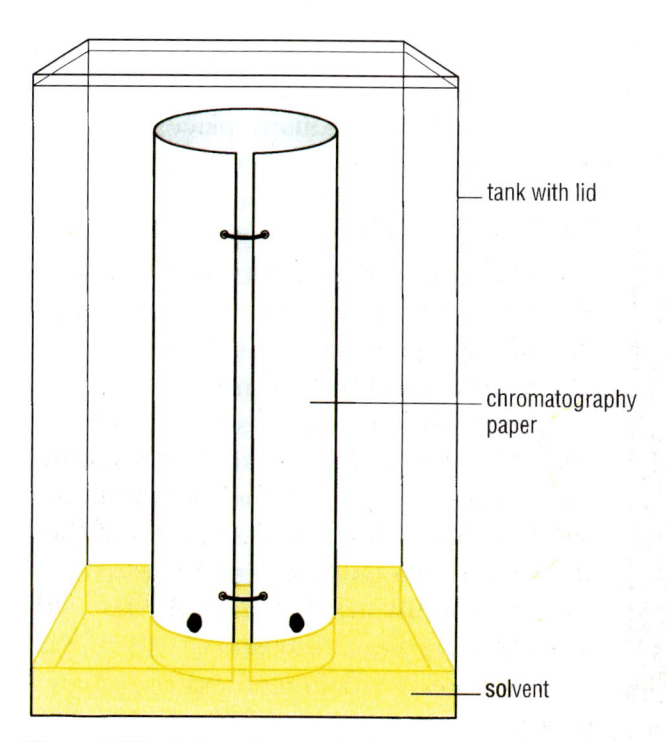

Figure 5.16 A chromatography tank.

MIXTURES AND SEPARATING TECHNIQUES

Separating a solvent from a solute/solvent mixture

During evaporation or boiling, the liquid solvent is lost to the air. If the solvent is important it can be separated from the mixture using a process called distillation.

In a very simple form of distillation, the solution is placed in a test-tube which is set up as shown in Figure 5.17.

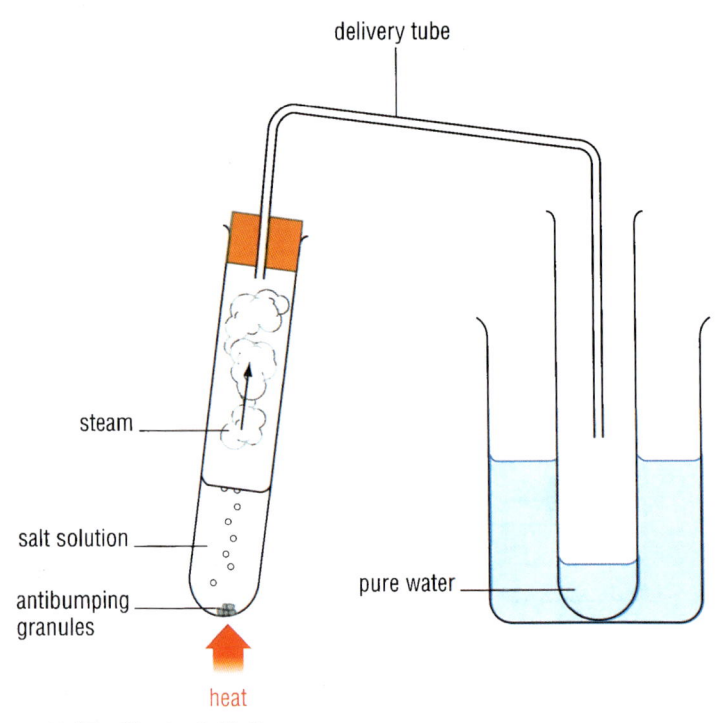

Figure 5.17 Simple distillation.

The antibumping granules provide many places where bubbles of gas may form as the water boils. The bubbles are small and steadily rise to the liquid surface where they burst. Without the granules, fewer but larger bubbles form that rise and burst with such force that they shake the test-tube.

As the water boils the steam moves along the delivery tube. At first the tube is cool enough to make some of the steam condense but as more steam passes along the tube it becomes hotter and no more condensation takes place. The cold water in the beaker keeps the walls of the second test-tube cool so that most of the steam condenses there and water collects at the bottom of the tube. The solid solute remains in the first tube. (Liquid and gas solutes are separated by fractional distillation.) The purity of the water can be checked by boiling it and recording its boiling point with a thermometer.

MIXTURES AND SEPARATING TECHNIQUES

Distillation with a Liebig condenser

The Liebig condenser is a glass tube surrounded by a glass chamber called a water jacket. During the distillation process, water is allowed to flow from the cold tap through the water jacket and down the sink. The water takes away the heat from the hot vapour in the tube of the condenser and causes it to condense. The liquid formed by the condensed vapour is called the distillate. It flows down the tube and drips into the collection flask.

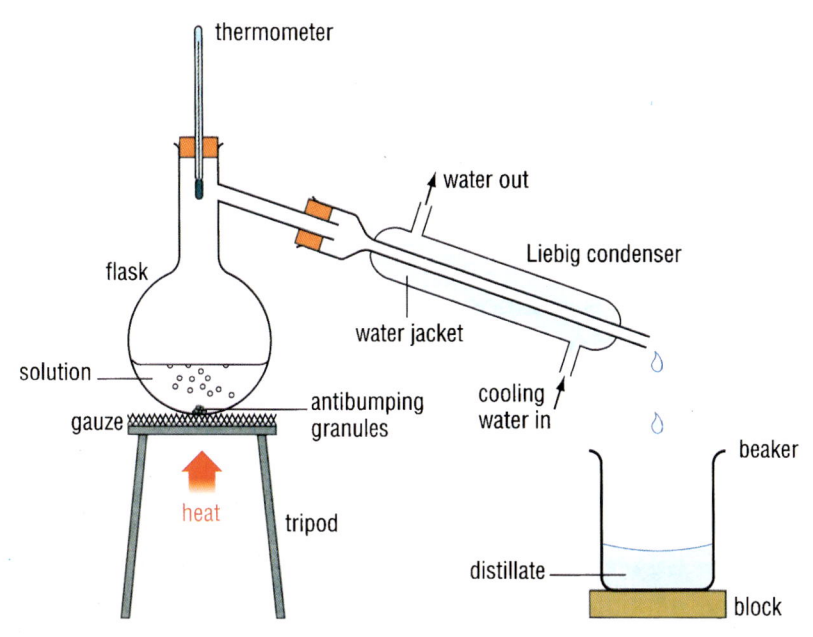

10 Why is the Liebig condenser more efficient than the simple distillation apparatus?

Figure 5.18 Distillation with a Liebig condenser.

Separating two immiscible liquids

When two immiscible liquids are mixed together they eventually form layers, if left to stand. This can be seen when oil and vinegar are mixed together to form salad dressing.

Figure 5.19 Salad dressing mixture after shaking (left) and after standing for 10 minutes (right).

73

■ MIXTURES AND SEPARATING TECHNIQUES

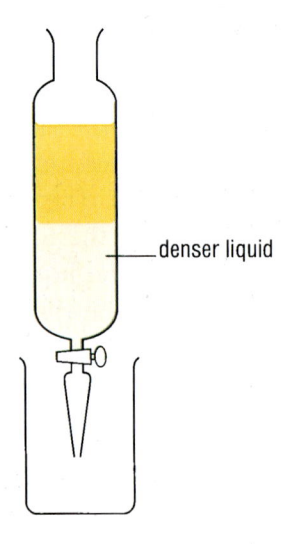

The less dense liquid forms a layer above the more dense liquid. The separating funnel (see Figure 5.20) can be used to separate them. The tap is opened to let the liquid in the lower layer flow away into a beaker. A second beaker can be used to collect the liquid from the upper layer.

denser liquid

Figure 5.20 A separating funnel.

◆ SUMMARY ◆

- A mixture is composed of two or more substances. There are no fixed amounts in which they combine. They can be separated by physically removing one substance from the other (*see page 62*).
- There is a wide range of mixtures in which solids, liquids and gases combine together (*see pages 62–63*).
- A solution is composed of a solute and a solvent (*see page 64*).
- When no more solute will dissolve in a solvent, the solution is said to be saturated (*see page 65*).
- The maximum mass of solute that will dissolve in 100 g of solvent, at a particular temperature, is known as the solubility of the substance at that temperature (*see page 65*).
- More gas will dissolve in cold water than in warm water (*see page 66*).
- There are many solvents. Substances that dissolve in one solvent may not dissolve in others (*see page 66*).
- Solids may be separated from each other with a sieve, magnetic separator or flotation cell (*see pages 66–67*).
- Solids may be separated from liquids by decanting, filtration or by using a centrifuge (*see pages 68–69*).
- A solid solute may be separated from a solvent by evaporation, crystallisation or chromatography (*see pages 70–71*).
- A solvent may be separated from a solid solute by distillation (*see page 72*).
- Two immiscible liquids can be separated by using a separating funnel (*see page 73*).

End of chapter question

How would you separate the different parts of a mixture of sand and salty water?

6 Atoms and elements

Elements and compounds

One of the main activities in chemistry is breaking down substances to discover what they are made of. During the course of this work chemists have discovered that some substances will not break down into simpler substances. These substances are called elements.

Discovery of the elements

Before 1669 the following elements had already been discovered – carbon, sulphur, iron, copper, arsenic, silver, tin, antimony, gold, mercury and lead. Some had been known for thousands of years, although they had not been recognised as elements. The order in which the other elements were discovered is shown in Table 6.1. This table uses mainly European historical data but it is known that the Chinese and people in Muslim countries also practised alchemy, so some of the elements could have been discovered by them at an earlier date.

1 How many elements were discovered in:
   a) the 17th Century or earlier,
   b) the 18th Century,
   c) the 19th Century?
2 Which three scientists discovered the most elements?
3 How many Swedish scientists discovered new elements?
4 Which UK scientist discovered the most elements?

Table 6.1 The discovery of the elements.

Date	Element	Discoverer	Brief description
1669	Phosphorus	H. Brand (Germany)	white, red, black solid
1737	Cobalt	G. Brandt (Sweden)	reddish metal
1746	Zinc	A.S. Marggraf (Germany)	blue–white metal
1748	Platinum	A. de Ulloa (Spain)	blue–white metal
1751	Nickel	A.F. Cronstedt (Sweden)	silver–white metal
1753	Bismuth	C.F. Geoffroy (France)	silver–red metal
1766	Hydrogen	H. Cavendish (UK)	colourless gas
1771–1774	Oxygen	C.W. Scheele (Sweden) J. Priestley (UK)	colourless gas
1772	Nitrogen	D. Rutherford (UK)	colourless gas
1774	Chlorine	C.W. Scheele (Sweden)	green–yellow gas
1774	Manganese	J.G. Gahn (Sweden)	red–white metal
1781	Molybdenum	P.J. Hjelm (Sweden)	silver–grey metal
1783	Tellurium	F.J. Muller (Austria)	silver–grey solid
1783	Tungsten	J.J. de Elhuya, F. de Elhuya (Spain)	grey metal
1789	Zirconium	M.H. Klaproth (Germany)	shiny, white metal
1789	Uranium	M.H. Klaproth (Germany)	blue–white metal
1794	Yttrium	J. Gadolin (Finland)	shiny, grey metal
			(continued)

ATOMS AND ELEMENTS

Date	Element	Discoverer	Brief description
1795	Titanium	M.H. Klaproth (Germany)	silvery metal
1798	Beryllium	N-L Vauquelin (France)	brown powder
1798	Chromium	N-L Vauquelin (France)	silvery metal
1801	Niobium	C. Hatchett (UK)	grey metal
1802	Tantalum	A.G. Ekeberg (Sweden)	silvery metal
1803	Cerium	J.J. Berzelius, W. Hisinger (Sweden) M.H. Klaproth (Germany)	grey metal
1803	Palladium	W.H. Wollaston (UK)	silver–white metal
1804	Rhodium	W.H. Wollaston (UK)	grey–blue metal
1804	Osmium	S. Tennant (UK)	blue–grey metal
1804	Iridium	S. Tennant (UK)	silver–white metal
1807	Potassium	H. Davy (UK)	silver–white metal
1807	Sodium	H. Davy (UK)	silver–white metal
1808	Magnesium	H. Davy (UK)	silver–white metal
1808	Calcium	H. Davy (UK)	silver–white metal
1808	Strontium	H. Davy (UK)	silver–white metal
1808	Barium	H. Davy (UK)	silver–white metal
1808	Boron	J. Gay-Lussac, L. Thernard (France)	dark brown powder
1811	Iodine	B. Courtois (France)	grey–black solid
1817	Lithium	J.A. Arfwedson (Sweden)	silver–white metal
1817	Cadmium	F. Stromeyer (Germany)	blue–white metal
1818	Selenium	J.J. Berzelius (Sweden)	grey solid
1824	Silicon	J.J. Berzelius (Sweden)	grey solid
1825–1827	Aluminium	H.C. Oersted (Denmark) F. Wohler (Germany)	silver–white metal
1826	Bromine	A.J. Balard (France)	red–brown liquid
1829	Thorium	J.J. Berzelius (Sweden)	grey metal
1830	Vanadium	N.G. Sefstrom (Sweden)	silver–grey metal
1839	Lanthanum	C.G. Mosander (Sweden)	metallic solid
1843	Terbium	C.G. Mosander (Sweden)	silvery metal
1843	Erbium	C.G. Mosander (Sweden)	silver–grey metal
1844	Ruthenium	K.K. Klaus (Estonia)	blue–white metal
1860	Caesium	R.W. Bunsen, G.R. Kirchhoff (Germany)	silver–white metal
1861	Rubidium	R.W. Bunsen, G.R. Kirchhoff (Germany)	silver–white metal
1861	Thallium	W. Crookes (UK)	blue–grey metal
1863	Indium	F. Reich, H.T. Richter (Germany)	blue–silver metal

(continued)

ATOMS AND ELEMENTS

Date	Element	Discoverer	Brief description
1868	Helium	J.N. Lockyer (UK)	colourless gas
1875	Gallium	L. de Boisbaudran (France)	grey metal
1878	Ytterbium	J-C-G de Marignac (Switzerland)	silvery metal
1878–1879	Holmium	J.L. Soret (France) P.T. Cleve (Sweden)	silvery metal
1879	Scandium	L.F. Nilson (Sweden)	metallic solid
1879	Samarium	L. de Boisbaudran (France)	light grey metal
1879	Thulium	P.T. Cleve (Sweden)	metallic solid
1880	Gadolinium	J-C-G de Marignac (Switzerland)	silver–white metal
1885	Neodymium	C. Auer von Welsbach (Austria)	yellow–white metal
1885	Praseodymium	C. Auer von Welsbach (Austria)	silver–white metal
1886	Dysprosium	L. de Boisbaudran (France)	metallic solid
1886	Fluorine	H. Moissan (France)	green–yellow gas
1886	Germanium	C.A. Winkler (Germany)	grey–white metal
1894	Argon	W. Ramsay, Lord Rayleigh (UK)	colourless gas
1898	Krypton	W. Ramsay, M.W. Travers (UK)	colourless gas
1898	Neon	W. Ramsay, M.W. Travers (UK)	colourless gas
1898	Polonium	Mme M.S. Curie (Poland/France)	metallic solid
1898	Xenon	W. Ramsay, M.W. Travers (UK)	colourless gas
1898	Radium	P. Curie (France), Mme M.S. Curie (Poland/France), M.G. Bermont (France)	silvery metal
1899	Actinium	A. Debierne (France)	metallic solid
1900	Radon	F.E. Dorn (Germany)	colourless gas
1901	Europium	E.A. Demarçay (France)	grey metal
1907	Lutetium	G. Urbain (France)	metallic solid
1917	Protactinium	O. Hahn (Germany), Fr L. Meitner (Austria), F. Soddy, J.A. Cranston (UK)	silvery metal
1923	Hafnium	D. Coster (Netherlands) G.C. de Hevesy (Hungary/Sweden)	grey metal
1925	Rhenium	W. Noddack, Fr I. Tacke, O. Berg (Germany)	white–grey metal
1937	Technetium	C. Perrier (France) E. Segre (Italy/USA)	silver–grey metal
1939	Francium	Mlle M. Percy (France)	metallic solid
1940	Astatine	D.R. Corson, K.R. Mackenzie (USA) E. Segre (Italy/USA)	metallic solid
1945	Promethium	J. Marinsky, L.E. Glendenin, C.O. Corgell (USA)	metallic solid

ATOMS AND ELEMENTS

Properties of elements and compounds

Only a very few of the substances you see around you are elements. The most common solid elements are metals such as aluminium and copper, though objects made of the elements gold and silver may be more obvious.

There are only two elements that are liquid at room temperature and standard pressure. They are mercury and bromine. Eleven elements are gases under normal conditions. Oxygen and nitrogen, which together form about 98% of the air, are two of them.

Each element has its own special properties. For example, sodium is a soft, silvery–white metal with a melting point of 97.86 °C and a boiling point of 884 °C and chlorine is a yellow–green gas with a melting point of −100.97 °C and a boiling point of −34.03 °C.

Most substances are made from two or more elements that are joined together. These substances are called compounds. They have properties which are different from the elements that make them. Common salt, for example, is a compound of sodium and chlorine and is a white solid with a melting point of 801 °C and a boiling point of 1420 °C. It easily forms crystals.

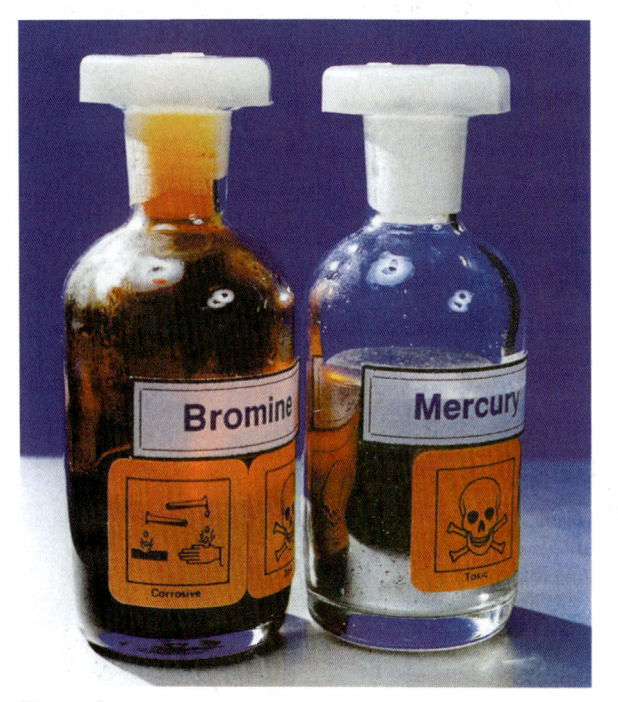

Figure 6.1 Mercury and bromine are liquid at room temperature.

Chemical symbols

Alchemists investigated materials in an attempt to find a way to make gold or a medicine which would extend the human life span. They wrote down details of their investigations using symbols to represent the substances they used or produced. The use of symbols saved them time. Figure 6.2 shows a few of the alchemists' symbols.

Many of the substances had been given a number of different names by different alchemists. When chemists began their work they used the alchemists' names, but this soon led to confusion.

It was decided that each substance used in an investigation or produced from it should be clearly identified by one name only so that reports of investigations could be clearly understood.

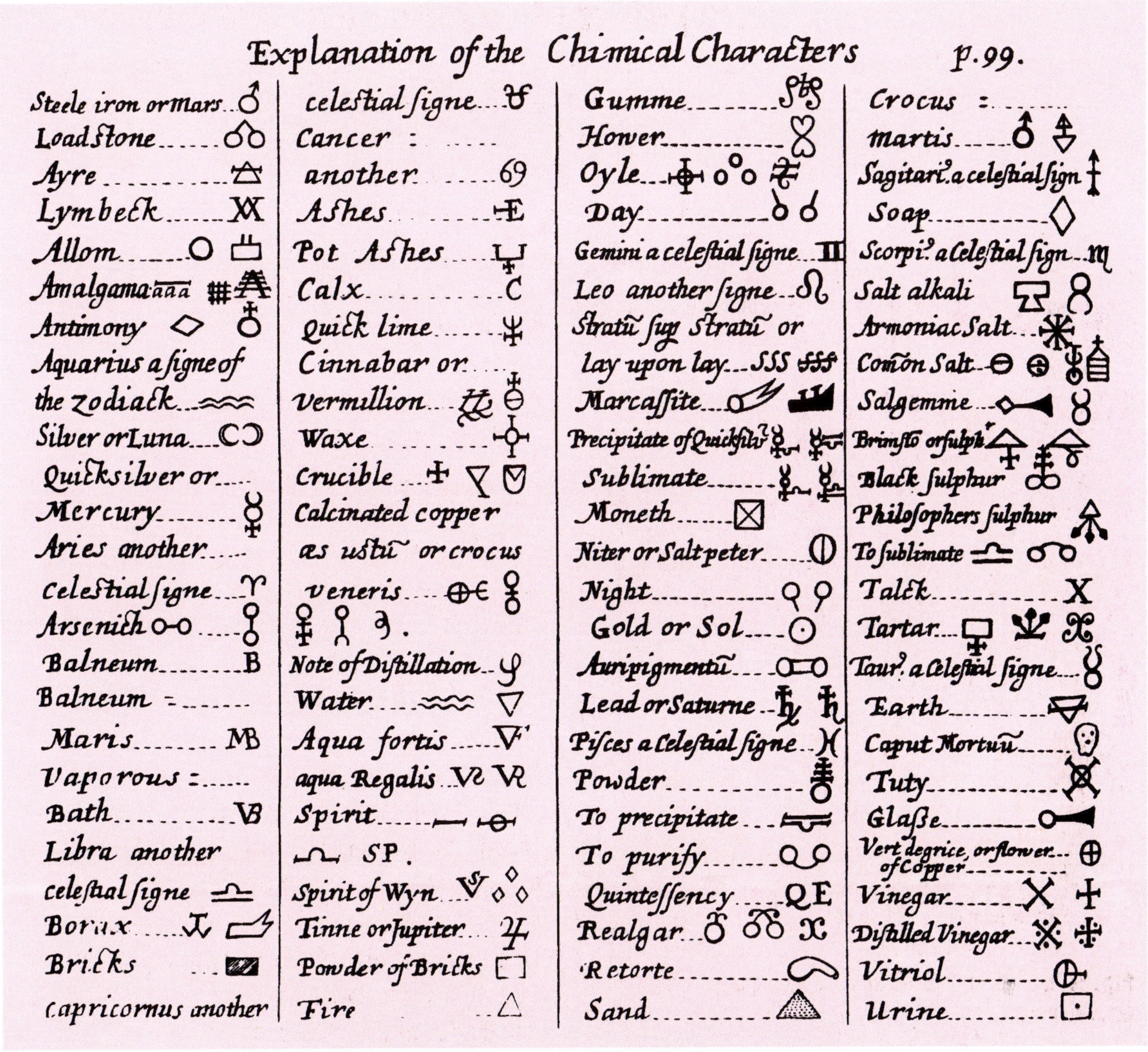

ATOMS AND ELEMENTS

Figure 6.2 Alchemists' symbols.

In 1787 Lavoisier and three other scientists set out the names of all the substances used in chemical investigations in a 300-page book.

In 1813 Jöns Jakob Berzelius introduced the symbols we still use to represent the elements. Each element was identified by the first letter of its name. If two or more elements began with the same letter another letter in the name was also used.

Some of the symbols are made from old names for the elements. Iron, for example, had an old name of ferrum and the symbol Fe is made from it. Silver was known as argentum and its symbol is Ag.

Sodium is known as natrium, and potassium is known as kalium in Latin and some other languages, and their symbols have been made from these names. The symbol for sodium is Na and the symbol for potassium is K.

ATOMS AND ELEMENTS

5 Why do some elements have two letters for their chemical symbol and others have only one?
6 Why isn't the symbol for silver S, and the symbol for potassium P?
7 How did some elements get their names?

The elements have received their names from a variety of sources. Some elements such as chlorine (from the Greek word meaning green colour) and bromine (from the Greek word for stench) are named after their properties. Other elements are named after places. The places may be as small as a village – strontium is named after Strontian in Scotland – or as large as a planet – uranium is named after the planet Uranus. A few elements, such as einsteinium, are named after people.

Atoms

Each element is made of atoms. An atom is about a ten-millionth of a millimetre across. It is made of sub-atomic particles (see Figure 6.3). At the centre of the atom is the nucleus. It is made from two kinds of sub-atomic particles called protons and neutrons. (Hydrogen is an exception because it has only a proton in its nucleus.) A proton has the same mass as a neutron. It also has a positive electrical charge, while a neutron does not have an electrical charge.

Around the nucleus are sub-atomic particles called electrons. Each electron has a negative electrical charge and travels at about the speed of light as it moves around the nucleus.

The number of electrons around the nucleus of an atom is the same as the number of protons in the

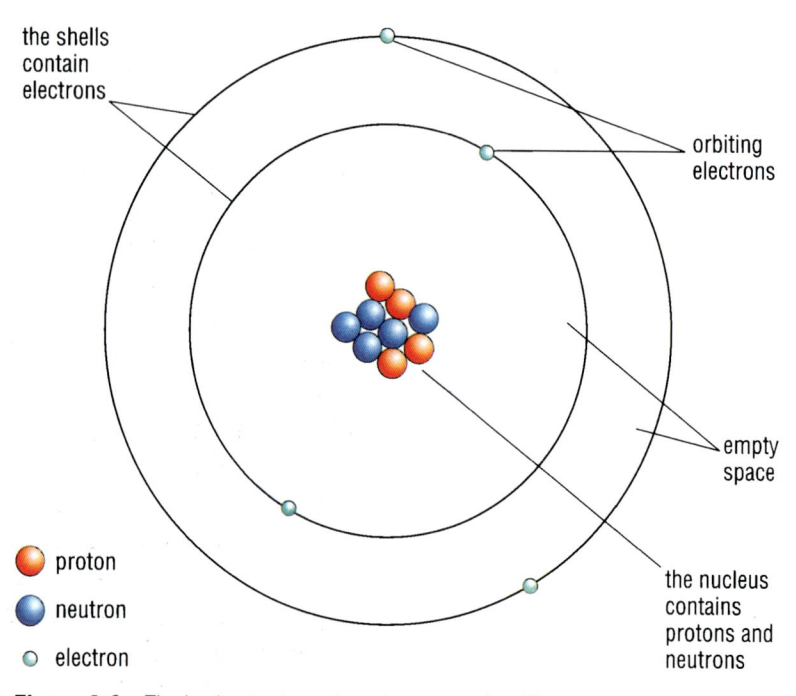

Figure 6.3 The basic structure of an atom, e.g. a beryllium atom.

ATOMS AND ELEMENTS

nucleus. The negative electrical charges on the electrons are balanced by the positive electrical charges on the protons. This balancing of the charges makes the atom electrically neutral – it has no electrical charge.

The fourth state of matter

Inside stars the temperature is so high that the atoms break up. Some of the electrons break away from the rest of the particles in the atom. The remaining particles form an electrically charged structure called an ion. This mixture of electrons and ions in a star is called plasma.

Plasma can also be made on the Earth by using low pressures in a glass container. If the container has an electrically charged rod at its centre the plasma will carry electricity to the glass. The path of electricity is shown by a flash of light through the plasma.

1 How is plasma different from other states of matter?

Figure A Plasma carrying electricity in a glass container.

The 20 lightest elements

All atoms have mass. The amount of mass of an atom is a measure of the amount of matter in it. An atom with a large number of protons, neutrons and electrons, like a lead atom (see page 162), has a greater mass than an atom with a small number of sub-atomic particles such as beryllium (see page 80).

The weight of an atom is the pull of the Earth's gravity on the atom and is related to the mass of the atom. The greater the mass of the atom the greater its weight. Atoms can then be arranged in order of their weights. Table 6.2 shows the 20 lightest elements, beginning with the lightest, hydrogen, and listed in order of increasing weight. You can find out when most of the

ATOMS AND ELEMENTS

elements were discovered by looking at Table 6.1 on pages 75–77 of this book. Carbon and sulphur were discovered before the time covered by the table.

Table 6.2 The chemical symbols of the 20 lightest elements (sodium follows on from neon).

Name	Symbol	Name	Symbol
Hydrogen	H	Sodium	Na
Helium	He	Magnesium	Mg
Lithium	Li	Aluminium	Al
Beryllium	Be	Silicon	Si
Boron	B	Phosphorus	P
Carbon	C	Sulphur	S
Nitrogen	N	Chlorine	Cl
Oxygen	O	Argon	Ar
Fluorine	F	Potassium	K
Neon	Ne	Calcium	Ca

The elements are sorted out and arranged in a table called the periodic table (see Figure 6.4). Chapter 15 covers the periodic table in greater detail.

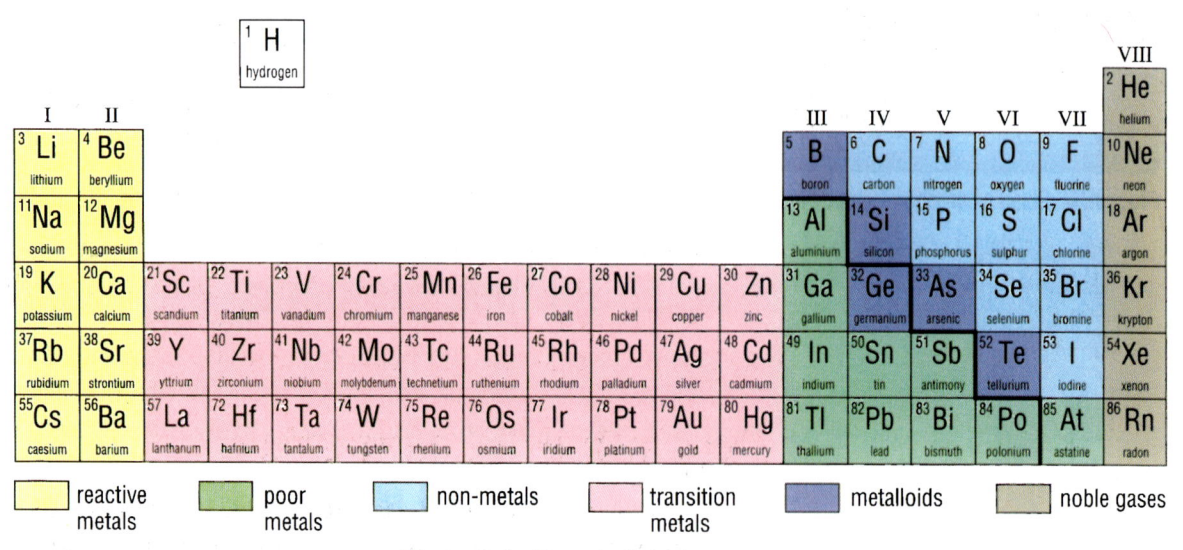

Figure 6.4 The periodic table.

Flame tests

Some metallic elements can be identified by the colour of the flame they produce when they are heated. The elements are tested in the following way.

ATOMS AND ELEMENTS

A nichrome wire is dipped in dilute hydrochloric acid and then dipped into a solid salt (see page 85) of the metal. The end of the wire is then held in a Bunsen burner flame and a coloured flame develops around it.

Metals and their flames

- Sodium: yellow.
- Potassium: lilac.
- Calcium: red.
- Copper: greeny blue.

Simple compounds

A compound is formed when the atoms of different elements join together. Water is a very common compound. It can be produced by burning hydrogen in oxygen. The word equation for this reaction is:

hydrogen + oxygen → water

Hydrogen and oxygen are both gases. They are in the same state when they react. Substances do not have to be in the same state to react to form compounds. For example, carbon in a block of glowing charcoal on a barbecue combines with oxygen in the air to make carbon dioxide. The word equation for this reaction is:

carbon + oxygen → carbon dioxide

When magnesium is heated in air, the oxygen in the air combines with the magnesium to form magnesium oxide:

magnesium + oxygen → magnesium oxide

During this reaction large amounts of heat and light are released. This synthesis reaction is a major feature of all firework displays (see Figure 6.5 and also page 3 where a Chinese firecracker firework is shown).

Figure 6.5 A firework display.

Proportions of elements

In a compound the elements are always present in the same proportions. For example, in iron sulphide there is always one atom of iron joined with one atom of sulphur. Two atoms of iron do not sometimes join to one atom of sulphur or one atom of iron sometimes join with three atoms of sulphur. The proportion of one element to the other in a compound is always the same. The elements in a compound are said to occur in fixed proportions.

ATOMS AND ELEMENTS

A mixture may be made up of elements (for example iron and sulphur), or compounds (for example iron sulphide and magnesium oxide), or even a mixture of the two (for example magnesium oxide and sulphur). Whatever the substances in a mixture, the proportions can vary widely (see page 62) – the proportions are not fixed.

♦ SUMMARY ♦

- Substances can be broken down into elements (*see page 75*).
- There is a chemical symbol for each element (*see page 78*).
- Each element is composed of atoms (*see page 80*).
- An atom contains protons, neutrons and electrons (*see page 80*).
- The 20 lightest elements can be arranged in order of their weight (*see page 82*).
- Some metallic elements can be identified by the colour of their flames (*see page 82*).
- A simple compound is formed when atoms of different elements join together (*see page 83*).
- In a compound, the elements are always present in the same proportions (*see page 83*).

End of chapter questions

1. Which elements are gases at normal temperatures as described in Table 6.1 page 75?
2. When the structure of the atom was discovered, some people compared it to the arrangement of the Sun and the planets in the Solar System. Use other resources, such as books and the internet, to examine the structure of the Solar System. Does it match the structure of an atom? Explain your answer.

7 Further reactions

In Chapter 6 we saw how the atoms of different elements took part in chemical reactions to make compounds. In this chapter we look at different kinds of chemical reactions and the compounds that they make. These reactions involve the neutralisation of acids, and the action of acids on metals, metal oxides and carbonates. We will also be looking at tests for common gases and reactions involving hydroxides and carbonates.

Salts

Some people have difficulty thinking about the word salt. All they can think about is sodium chloride – common salt. There are many different salts and some of them are useful. For example, potassium nitrate is used as a fertiliser, as a preservative of meat products and in making gunpowder. Magnesium sulphate is used as a laxative to ease constipation. Sodium stearate is a salt used to make soap.

A salt is formed as a result of a neutralisation reaction between an acid and a base (see page 25).

Preparing crystals of sodium chloride

One of the most widely used chemical compounds is common salt. It is the ingredient added to food to bring out its taste. It also has many uses in the chemical industry. Common salt is made from atoms of the elements sodium and chlorine.

A neutralisation reaction takes place between an acid and a base to produce a salt. This type of chemical reaction is used in the preparation of sodium chloride.

The point at which neutralisation takes place can be found by using universal indicator solution or a pH probe and by carefully measuring the volume of acid added to a known volume of alkali using a burette (see page 7). Here are the steps to take to prepare a salt by neutralisation (see Figure 7.1, overleaf).

1. Wear eye protection.
2. Use a measuring cylinder (see page 6) to measure out 25 cm³ of sodium hydroxide solution into a flask.
3. Place two drops of indicator in the sodium hydroxide solution and swirl it round so that the whole solution turns purple or the pH probe records a pH of 14.

FURTHER REACTIONS

1 Write the word equation for the reaction taking place in the neutralisation experiment.

2 Why does the experiment need to be repeated if you have used the universal indicator solution?

4 Place the flask under a burette containing hydrochloric acid. Add a small quantity of hydrochloric acid. Swirl the flask to mix the reactants and look for any colour change or change in the pH displayed by the pH probe.

5 Repeat step 4 until the colour of the indicator changes to green or the pH probe records a pH of 7. Neutralisation has occurred. Read off the volume of acid added from the burette.

6 If you have used the indicator solution repeat step 2 and then add the volume of acid you now know to be needed for neutralisation to take place.

7 Warm the neutralised solution to let the water evaporate leaving the salt crystals behind.

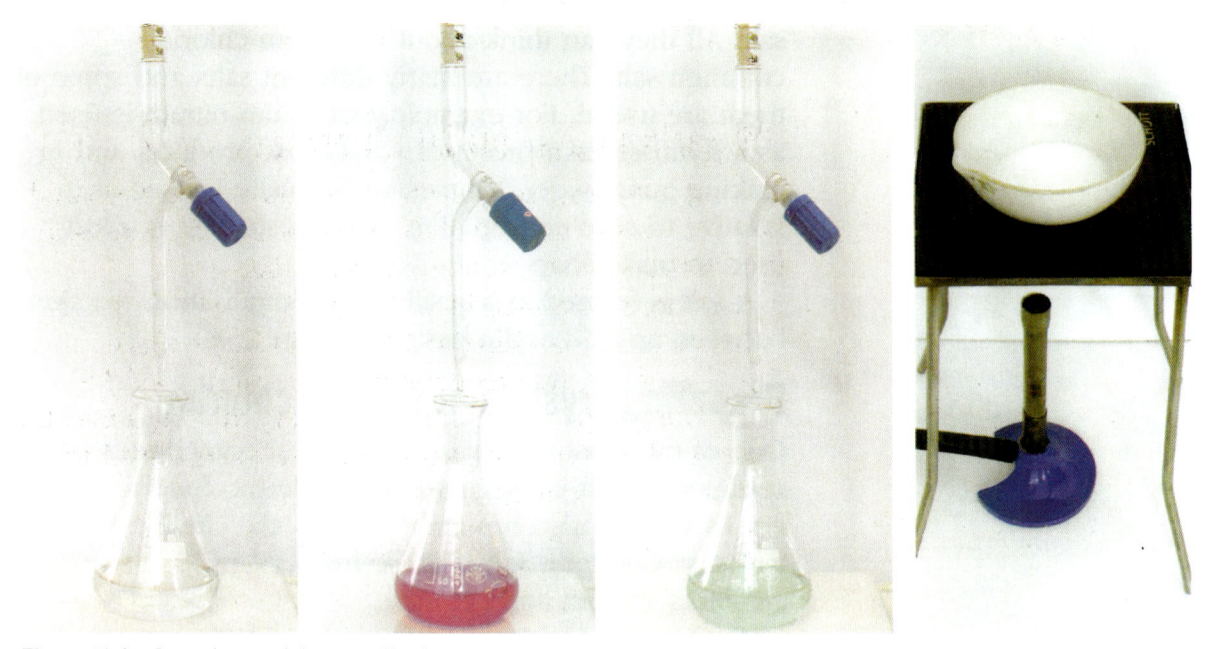

Figure 7.1 Preparing a salt by neutralisation.

The production of common salt

The neutralisation reaction is not used to produce the common salt we use in our food. Sodium chloride is a soluble substance and readily forms a solution with water. Sodium chloride present in rocks is dissolved by water passing through the rocks and is carried away in streams and rivers to the sea. The saltiness of sea water is due to the presence of sodium chloride.

In the past, ancient seas dried up and left a deposit of salt on the seabed. In time this deposit was covered with other rock and formed the mineral halite. This is also called rock salt.

In certain parts of the world, such as in England, halite is mined to provide salt. Along the coasts of some countries in warm parts of the world sea water is trapped in shallow pools. As the water evaporates in hot weather the salt is left behind and is collected.

Figure A Halite.

(continued)

FURTHER REACTIONS

Figure B A salt pan on a tropical coast.

Sea water is not a pure solution of sodium chloride. It contains other salts such as magnesium chloride.

Preparing other metal chlorides

Chlorides of zinc, magnesium and iron can be prepared by reacting the metal with hydrochloric acid. Here is a description of the reaction between zinc and hydrochloric acid (see Figure 7.2).

There is zinc and hydrochloric acid in the flask. Bubbles of gas are emerging from the surface of the acid. The gas passes along the delivery tube and into the boiling tube. Here the gas pushes the water out of the way and in time may fill the tube.

3 How could you test the gas in the test tube in Figure 7.2 to see if it was hydrogen?

Figure 7.2 Apparatus for the collection of hydrogen given off during the preparation of zinc chloride.

FURTHER REACTIONS

4 What are the word equations for the reaction of magnesium and iron with hydrochloric acid?

Figure 7.3 Zinc chloride.

5 What are the chemical reactions that take place in the making of sulphuric acid?

6 What are the physical changes that take place in the making of sulphuric acid?

The word equation for this reaction is:

zinc + hydrochloric acid → zinc chloride + hydrogen

While it is easy to see the hydrogen gas as it forms bubbles and pushes water out of the way in the boiling tube, the zinc chloride cannot be seen. The reason for this is that zinc chloride is soluble. A solution of zinc chloride is forming in the flask. When the reaction is complete, the solution can be warmed so that evaporation takes place (see Figure 7.3). The water changes to water vapour and escapes into the air and the zinc chloride is left behind.

Sulphuric acid

You may know a little about sulphuric acid if you have studied acid rain and its effects on the environment. Power stations release large amounts of sulphur dioxide into the atmosphere as they burn fossil fuels. The sulphur dioxide reacts with water in rain drops to form dilute sulphuric acid.

Sulphuric acid has many uses (see page 89) and is made in the chemicals industry in the following way.

Sulphur is melted and sprayed into a furnace where it combines with oxygen in the air to form sulphur dioxide gas. This gas and more air then pass over a catalyst (see page 156) of vanadium pentoxide. This speeds up the reaction between sulphur dioxide and oxygen in the air to form sulphur trioxide. If this gas was added to water, a very violent reaction would take place which would be difficult to control so it is dissolved in some previously made sulphuric acid. This reaction is much easier to control. When the sulphur trioxide dissolves in

Figure 7.4 There is a sulphuric acid plant in the centre of this photograph.

FURTHER REACTIONS

the sulphuric acid it makes a very concentrated acid solution called oleum. This is diluted with water to reduce the concentration of the acid so that it can be used in industry and school laboratories.

> **For discussion**
>
> Sulphur dioxide is a strong smelling, harmful gas. Concentrated sulphuric acid is corrosive and removes water from biological substances such as sugar to leave just carbon behind (see Figure 7.5).
>
> What fears might people have about a sulphuric acid plant being built close to their homes (see Figure 7.4)? How could the people be reassured that it was safe? What would be the cost–benefit analysis of setting up a sulphuric acid plant near a town?

Figure 7.5 This piece of sugar has been dehydrated by sulphuric acid.

Uses of sulphuric acid

The ways in which sulphuric acid affects our lives may be considered in more detail by taking an everyday object such as a can of vegetable soup. The vegetable plants to make the soup were grown with the aid of fertilisers. Sulphuric acid is used to make a type of fertiliser called ammonium sulphate and it is also used to make calcium sulphate – a component of a fertiliser called superphosphate.

The vegetables have been processed in a factory where detergents are used to clean the working surfaces and utensils used to make the soup. Detergents are made from reactions which take place between sulphuric acid and chemicals extracted from oil.

The can of soup may have been delivered to the shop or supermarket in a white van. The colour is due to titanium, a metal extracted from its ore by using sulphuric acid. The lights and other electrical components on the van are powered from the battery which contains sulphuric acid.

> **For discussion**
>
> How could the fact that sulphuric acid is used in the making of fibres, soap and insecticide be included in the story of the can of soup?

Salts of sulphuric acid

The salts of sulphuric acid are called sulphates. They can be made by adding sulphuric acid to a metal, a metal oxide or a metal carbonate.

Reaction with a metal

Zinc forms a salt with sulphuric acid.

The word equation for the reaction between zinc and sulphuric acid is:

zinc + sulphuric acid → zinc sulphate + hydrogen.

FURTHER REACTIONS

Some useful sulphates

Calcium sulphate forms naturally in the Earth. It forms the mineral called gypsum. When gypsum is heated it forms plaster of Paris which is used to stop broken limbs from moving.

Two other sulphates are used in medicine. Magnesium sulphate is known as Epsom salts and is used to cure constipation. Barium sulphate is an unusual sulphate in that it does not dissolve in water. A thick suspension of a radio labelled barium sulphate is used as a barium 'meal'. Doctors can watch the barium sulphate pass through the intestines of a person who has drunk it by taking X-ray pictures.

Sodium sulphate is used to treat wood fibres in the paper-making industry and is also used to make glass.

Figure A This limb is protected by a plaster of Paris case.

> **For discussion**
>
> There is very little hydrogren produced when sulphuric acid is added to calcium. The salt that is produced, calcium sulphate, is insoluble. Use this information to explain why the reaction does not take place for long.

7 Write the word equation for the reaction between zinc oxide and hydrochloric acid.

8 Write a general word equation which describes the reaction between a metal oxide and an acid.

Reaction with a metal oxide

Copper oxide is a black powder but when sulphuric acid is added to it a blue solution forms. This indicates that a chemical reaction has taken place. The word equation for this reaction is:

copper oxide + sulphuric acid → copper sulphate + water

If there is too much copper oxide present some will remain at the bottom of the beaker. The unreacted copper oxide can be removed by filtration. The copper sulphate can be removed from the solution by allowing the water in the solution to evaporate (see Figure 7.7).

Figure 7.6 Copper oxide (left), sulphuric acid (centre), and copper sulphate (right).

FURTHER REACTIONS

Figure 7.7 Copper sulphate crystals.

Reaction with a metal carbonate

The apparatus shown in Figure 7.2 can also be used to investigate the reaction between a metal carbonate and an acid. However, the gas that is produced in this reaction is not hydrogen. It is carbon dioxide.

When sulphuric acid is added to calcium carbonate the following word equation describes the reaction:

calcium + sulphuric → calcium + water + carbon
carbonate   acid   sulphate   dioxide

Nitric acid

When lightning flashes (see Figure 7.8), the heat produced causes a chemical reaction between oxygen and nitrogen in the air. They combine and form oxides of nitrogen which dissolve in rainwater and form a very dilute solution of nitric acid. When this acid reaches the ground, it forms nitrates with some of the elements in the soil. These nitrates are used by plants for growth.

Figure 7.8 Lightning.

FURTHER REACTIONS

Nitric acid is also made in a chemical plant like sulphuric acid. Air and ammonia (see page 157) are the raw materials and a catalyst made of platinum and rhodium is used to speed up the reaction.

Salts of nitric acid

In the laboratory, nitrates can be made by adding nitric acid to alkalis, bases and carbonates. The word equations for the reactions are shown below.

Reaction with an alkali

sodium + nitric → sodium + water
hydroxide   acid   nitrate

Reaction with a base

magnesium + nitric → magnesium + water
oxide   acid   nitrate

Reaction with a carbonate

calcium + nitric → calcium + carbon + water
carbonate acid   nitrate   dioxide

Sodium nitrate, magnesium nitrate and calcium nitrate are salts formed from nitric acid.

From nitrogen to fertiliser

Nitrogen forms most of the air. In 1908 Fritz Haber (1868–1934), a German chemist, discovered a way of making nitrogen and hydrogen gases react together to make ammonia gas in the laboratory. By 1913 Carl Bosch (1874–1940), also a German chemist, had scaled up the laboratory apparatus. The ammonia gas was mixed with hot nitric acid and then sprayed into the top of a tower. Cold air was pumped into the bottom of the tower. As the droplets of mixture fell through the rising air, they cooled and formed pellets of ammonium nitrate.

When ammonium nitrate is put onto soil it breaks down into substances called minerals that plants take in through their roots. The plants extract the nitrogen from the minerals and use it to make proteins to form their cells. Fertilisers are added to many crops to help them produce as much food as possible.

Figure A Manufacturing ammonium nitrate.

FURTHER REACTIONS

Oxides and combustion

Test for oxygen

Oxygen causes substances to burn in it if they are hot enough. This property of oxygen can be used to test for it.

1 A splint is lit and then blown out.
2 The glowing end of the splint is plunged into a gas jar which contains an unknown gas.
3 If the gas is oxygen, the splint bursts into flame.

Combustion of carbon

The simplest combustion reaction is one between an element and oxygen. Charcoal is made from the element carbon. When it is heated in air and plunged into oxygen it burns with a red glow and a colourless gas is produced. If the gas is bubbled into lime water, the lime water turns cloudy showing that the gas is carbon dioxide. The word equation for the combustion of carbon in oxygen is:

carbon + oxygen → carbon dioxide

Test for carbon dioxide

Lime water is a dilute solution of calcium hydroxide. It is used to test for carbon dioxide gas. If a gas is thought to be carbon dioxide, it is bubbled through lime water. If carbon dioxide is present, a chemical reaction takes place in which calcium carbonate is made. This white substance is insoluble in water and forms a white precipitate which makes the lime water cloudy or milky (see Figure 7.9).

9 Why does a glowing splint burst into flame in a gas jar of oxygen but not in air? Air is a mixture of oxygen and other gases.

Clear lime water

Carbon dioxide is bubbled through

Lime water turns cloudy

Figure 7.9 A test for carbon dioxide.

FURTHER REACTIONS

Test for hydrogen

Hydrogen is a colourless gas that does not smell. It can be detected by the following test. Hold a small test-tube of hydrogen upside down and remove the bung. Hold a lighted splint below the open mouth of the test-tube and a popping sound will be heard. The hydrogen in the tube combines with the oxygen in the air and this explosive reaction makes the popping sound.

The word equation for the reaction is:

hydrogen + oxygen → hydrogen oxide

The common name for hydrogen oxide is water, and this is the word that is normally used in word equations:

hydrogen + oxygen → water

Combustion of metals

Metals are elements and when they burn in oxygen they, too, produce oxides.

Sodium

You can see how sodium burns in oxygen on page 137. The word equation for this reaction is:

sodium + oxygen → sodium oxide

Magnesium

The white light in a firework display (see page 83) is due to the combustion of magnesium. The word equation for this reaction is:

magnesium + oxygen → magnesium oxide

If a piece of magnesium is weighed before it is burnt and the metal oxide that is produced is also weighed, it will be found that the metal oxide weighs more than the metal. This is due to the oxygen atoms combining with the magnesium atoms to make the compound magnesium oxide.

A brief survey of hydroxides

Sodium hydroxide

The most widely used hydroxide in the chemistry laboratory is sodium hydroxide. It is made in the chemical industry by passing a current of electricity through a strong solution of sodium chloride.

Sodium hydroxide is an alkali (see page 23). It is corrosive and can damage the skin. Its common name is caustic soda.

FURTHER REACTIONS

sodium hydroxide

soap | paper | aluminium extraction | inorganic/organic sodium salts | effluent treatment | textiles (e.g. rayon, wool and cotton)

Figure 7.10 The uses of sodium hydroxide.

Calcium hydroxide

If calcium carbonate is heated strongly, it takes part in a decomposition reaction as the word equation describes:

calcium carbonate → calcium oxide + carbon dioxide

Calcium oxide is known as lime or quicklime. When water is added to it, the lime expands and heat is given out and calcium hydroxide called slaked lime is produced. As this reaction produces heat, it is known as an exothermic reaction (see page 144).

10 What is the word equation for the addition of water to lime?

Figure 7.11 Calcium hydroxide is used when tanning hides.

Coloured hydroxides

Some coloured hydroxides can be made by adding sodium hydroxide to solutions of metal salts. The colours of certain precipitates are used as tests to identify iron, copper and aluminium in chemical compounds whose composition is not known.

A white precipitate of aluminium hydroxide is made when sodium hydroxide is added to a solution of aluminium sulphate. The word equation for this reaction is:

aluminium sulphate + sodium hydroxide → aluminium hydroxide + sodium sulphate

FURTHER REACTIONS

11 Write a word equation for the reaction between copper sulphate solution and sodium hydroxide.

12 Write a word equation for the reaction between iron chloride solution and sodium hydroxide.

When sodium hydroxide is added to a copper sulphate solution a pale blue copper hydroxide is produced.

When sodium hydroxide is added to a solution of a salt of iron called iron chloride a brown substance called iron hydroxide is produced.

Two common carbonates

Calcium carbonate

Many regions of the world have large areas of landscape made from calcium carbonate. It forms rocks known as limestone.

Figure 7.12 Pinnacles Desert in Australia: a famous limestone landscape.

Heating limestone

A simple thermal decomposition can be performed by heating a small piece of limestone. Limestone is made of a compound called calcium carbonate. The heat breaks down the compound into calcium oxide and carbon dioxide. The word equation for this reaction is:

calcium → calcium + carbon
carbonate   oxide   dioxide

Calcium oxide does not break down when it is heated, but at very high temperatures it becomes incandescent and gives out a bright white light known as limelight. This was used to light the stages of theatres before electricity was available and gave rise to the expression 'in the limelight'.

Figure 7.13 The 'limelight man' in an old theatre.

96

FURTHER REACTIONS

Copper carbonate
Copper carbonate exists naturally as the mineral malachite.

Figure 7.14 Malachite.

When copper carbonate is heated it breaks down into copper oxide and carbon dioxide (see page 43).

Sodium hydrogencarbonate
This compound is also known by these common names: sodium bicarbonate or bicarbonate of soda. It is an alkali and when it is added to an acid, a neutralisation reaction takes place. The word equation for the reaction between sodium hydrogencarbonate and hydrochloric acid is:

sodium + hydrochloric → sodium + water + carbon
hydrogencarbonate   acid   chloride   dioxide

Sodium hydrogencarbonate is used in medicine to cure ailments caused by the stomach producing too much acid.

FURTHER REACTIONS

♦ SUMMARY ♦

- Sodium chloride (common salt) can be prepared by a neutralisation reaction (*see page 85*).
- Metal chlorides can be prepared by reacting some metals with hydrochloric acid (*see page 87*).
- Sulphuric acid is made in a chemical plant (*see page 88*).
- Sulphates can be prepared by reacting some metals, metal oxides or metal carbonates with sulphuric acid (*see pages 89–91*).
- Nitric acid is made when lightning occurs and in a chemical plant (*see page 91*).
- Nitric acid produces salts called nitrates when it reacts with alkalis, bases and carbonates (*see page 92*).
- A glowing splint is used to test for oxygen (*see page 93*).
- When carbon burns in oxygen, carbon dioxide is produced (*see page 93*).
- Lime water is used to test for carbon dioxide (*see page 93*).
- A lighted splint is used to test for hydrogen (*see page 94*).
- When metals burn in air they produce oxides (*see page 94*).
- Sodium hydroxide is the most widely used hydroxide in a chemistry laboratory (*see page 94*).
- Calcium hydroxide is also called slaked lime (*see page 95*).
- Some metal salts produce coloured hydroxides when sodium hydroxide is added to them (*see page 95*).
- Limestone is formed from calcium carbonate (*see page 96*).
- Copper carbonate exists naturally as malachite (*see page 97*).
- Sodium hydrogencarbonate is an alkali (*see page 97*).

End of chapter question

How useful are acids and alkalis? Explain your answer using information provided in this chapter. You may also use the internet to provide more details.

8 Compounds and mixtures

In the introduction to this book (see page 1) the formation of elements in stars was described. When stars form from clouds of gas and dust, planets may also form around them. The planets and their atmospheres are formed from elements and groups of elements called compounds.

Mixing elements

Each element has its own particular properties. Sulphur, for example, is yellow and if shaken with water it will tend to float. Iron is black and magnetic and produces hydrogen when it is placed in hydrochloric acid.
If the two elements are mixed together, a grey–black powder is produced. The colour depends on the amount of sulphur mixed with the iron. Although the two elements are close together, their properties do not change. If a magnet is stroked over the mixture, iron particles leap up and stick to it. If the mixture is shaken with water the sulphur will tend to float.

Figure 8.1 Black iron (left) and yellow sulphur (centre) mix to form a grey–black powder (right).

From elements to a compound

However, if the mixture of iron and sulphur is heated a change takes place. The atoms of iron and sulphur join together and form a compound called iron sulphide. It does not have the yellow colour of the sulphur or the magnetic properties of the iron. It has its own properties – it is a black non-magnetic solid. All compounds have properties which differ from the elements that formed them.

COMPOUNDS AND MIXTURES

Figure 8.2 The formation of iron sulphide.

The reaction which takes place when iron and sulphur are heated is shown in the word equation:

iron + sulphur → iron sulphide

Synthesis reactions

When two or more substances take part in a chemical reaction to make one compound the reaction is called a synthesis. The reaction of iron and sulphur is an example.

When a compound forms it may be in a different state to the elements from which it formed. For example, when water forms from hydrogen and oxygen the two gases produce a liquid. The word equation for this reaction is:

hydrogen + oxygen → water

Solutions

A solution is a mixture of a solute (a substance that dissolves in a solvent) and a solvent (substance in which a solute dissolves). You can find out more about solutions on page 64. Two important physical properties of solutions are freezing points and boiling points.

Freezing point

A pure solvent freezes at a certain temperature. When a solute is added to it and a solution is made, the solution freezes at a lower temperature than the solvent. The freezing point is said to have been depressed. If more solute is added to the solution, the freezing point is depressed further.

If you live in a country where ice forms on the roads in winter, you will know that the road is treated with salt and sand. The salt mixes with the ice and lowers its freezing point. This makes the ice melt. The sand simply increases friction between tyres and the road and prevents them skidding.

1 Draw a graph to show how you think the freezing point changes as more and more solute is added to a solution.

COMPOUNDS AND MIXTURES

Figure 8.3 Salt is used to treat icy roads.

2 Would you expect sea water to boil at the same temperature as fresh water? Explain your answer.

Boiling point

A pure solvent boils at a certain temperature. When a solute is added to it, the solution produced boils at a higher temperature than the solvent. This raising of the boiling point is called boiling point elevation.

Separating compounds

If a mixture contains a soluble compound and an insoluble compound, the compounds can be separated in the following way.

1 The mixture is added to a solvent (usually water).
2 The mixture and solvent are stirred until one of the compounds has dissolved.
3 The insoluble compound and the solution are separated by filtration (see page 68).
4 The insoluble compound is set out on a paper towel and allowed to dry.
5 The solution is evaporated by heating gently until it becomes concentrated. Then it is left to cool. The concentrated solution continues to evaporate and in time crystals form (see page 70).

Precipitation

Some compounds dissolve in liquids to make solutions (see page 64). When some solutions are mixed together a precipitate is formed. A precipitate is tiny particles that do not dissolve in the mixture of the solutions. The precipitate makes the liquid cloudy.

101

COMPOUNDS AND MIXTURES

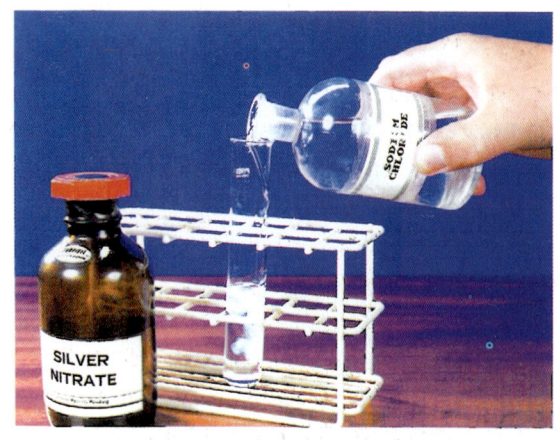

Figure 8.4 The precipitation of silver chloride.

If silver nitrate solution is poured into a solution of sodium chloride, a chemical reaction takes place which produces silver chloride. This forms a white precipitate.

silver + sodium → silver + sodium
nitrate   chloride   chloride   nitrate

Separating elements from compounds

Electricity can be used to separate the elements from a compound.

Electrolysis

Electrolysis is the decomposition of an electrolyte using electricity. An electrolyte is a solution or a molten solid through which the current passes. The elements which are produced as a result of the decomposition collect at the carbon rods. These rods are known as electrodes. Platinum is another material which is used to make electrodes. Carbon and platinum are used because they do not usually take part in chemical reactions with the electrolyte or the chemicals which form on their surfaces. One electrode has a positive charge and is called the anode. The other electrode has a negative charge and is called the cathode (see Figure 8.5).

3 What is an electrolyte?

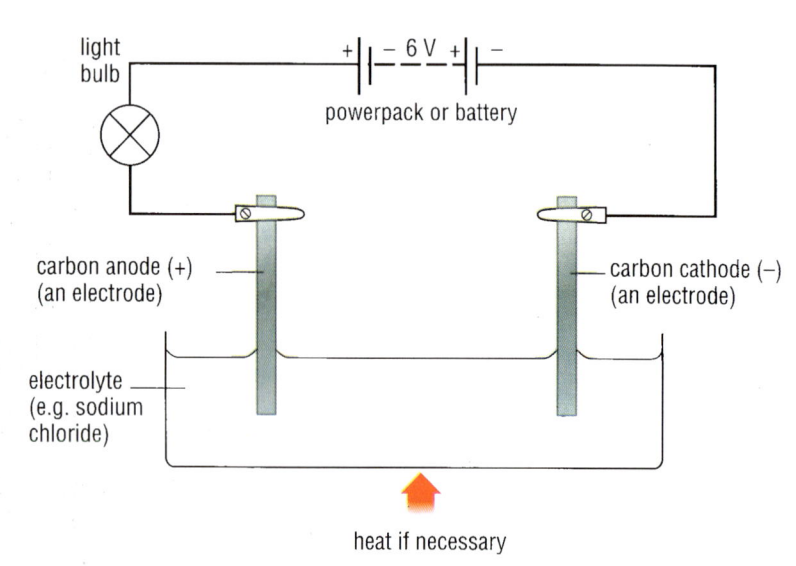

Figure 8.5 Apparatus for electrolysis.

COMPOUNDS AND MIXTURES

Pure molten electrolyte

When sodium chloride is molten and a current of electricity is passed through it, sodium metal is produced at the cathode and chlorine gas is produced at the anode. (The sodium metal is difficult to see.)

If lead bromide is heated until it melts and a current of electricity is passed through it, an amount of lead is produced at the cathode which is easier to see. Bromine gas is produced at the anode. This experiment should always be performed in a fume cupboard because the bromine gas and lead vapour that escape from the molten electrolyte are toxic.

The electrolysis of water

Pure water does not conduct electricity but if a small amount of sulphuric acid is added to it a current can pass. When the water is set up in a Hoffmann voltameter and the current is left to pass for some time, gas collects near the top of the apparatus. Hydrogen and oxygen collect at the top of the voltameter as Figure 8.7 shows. You can see that there is twice as much hydrogen produced as oxygen. This suggests that a molecule of water is made from two hydrogen atoms and one atom of oxygen.

From observations like this, chemists have worked out a chemical formula for different kinds of molecule. For example the chemical formula of water is H_2O.

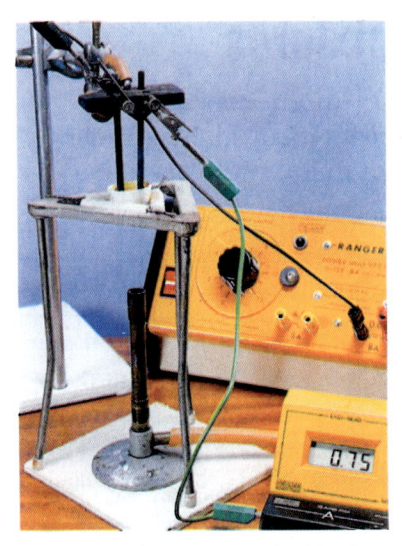

Figure 8.6 The electrolysis of lead bromide in a fume cupboard.

4 What does the formula for:
  a) sodium hydroxide NaOH, and
  b) sulphuric acid H_2SO_4 tell you about the atoms present in the molecules?

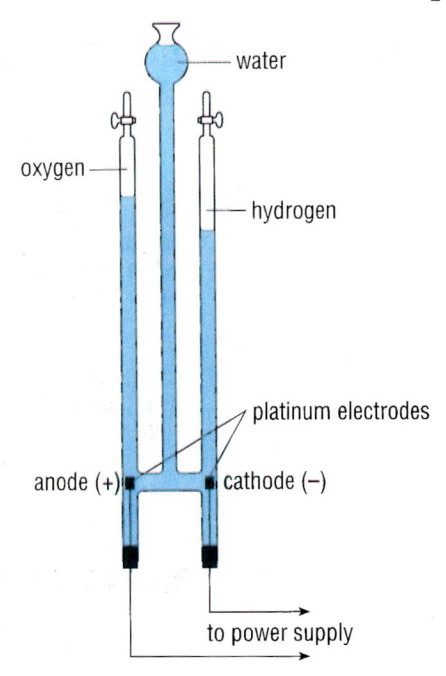

Figure 8.7 A Hoffmann voltameter.

COMPOUNDS AND MIXTURES

Separating liquid mixtures

Mixtures are made up of two or more compounds. Differences in their physical properties, such as boiling points, can be used to separate them.

Separating two miscible liquids

Fractional distillation

Two liquids with quite different boiling points, such as water (100 °C) and ethanol (78 °C), can be separated by fractional distillation.

The separation occurs in the fractionating column. This is filled with glass beads that provide a large surface area. During the fractionating process the liquids condense and evaporate from the surface many times. At first, in the lower part of the column, the water and ethanol vapours condense together on the cold beads. They warm them up and some of the ethanol and a little water vapour evaporate and move up a little further, then condense. As the solution continues to boil, this process of condensation and evaporation is repeated all the way up the column. Each time, more of the liquid with the lower boiling point – the ethanol – rises further, until it reaches the top. The ethanol vapour then passes down the Liebig condenser and is collected in the flask. As the vapour passes the thermometer, a temperature of 78 °C is recorded. This is the temperature at which pure ethanol boils. When most of the ethanol has passed the thermometer the temperature starts to rise. At this point, the process is stopped as the liquid in the flask is nearly all water.

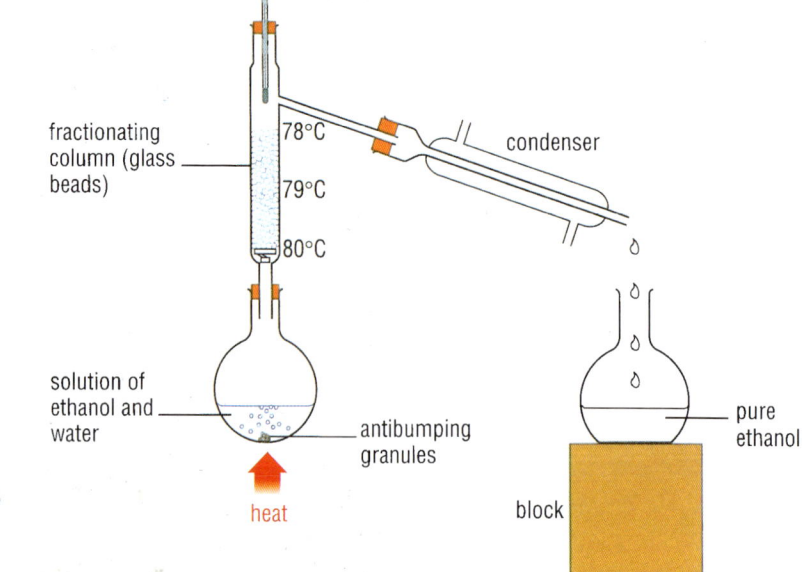

5 Why did the thermometer show 78 °C for some time?

6 If the distillation was left to run, what temperature would the thermometer rise to? Explain your answer.

7 How would you use the apparatus in Figure 8.8 to get a flask of ethanol and a flask of water?

Figure 8.8 The distillation of ethanol.

COMPOUNDS AND MIXTURES

Fractional distillation of liquid air
Air can be cooled until it becomes a liquid and then gently warmed to release its different gases.

The petrochemicals industry

Petrochemicals are made from petroleum and natural gas. Petroleum means rock oil. It is usually simply called oil.

Natural gas and oil are made from the dead bodies of tiny animals and plants that lived in the seas over 200 million years ago. The bodies decayed to form molecules called hydrocarbons. These are made of atoms of carbon and hydrogen as Figure 8.9 shows.

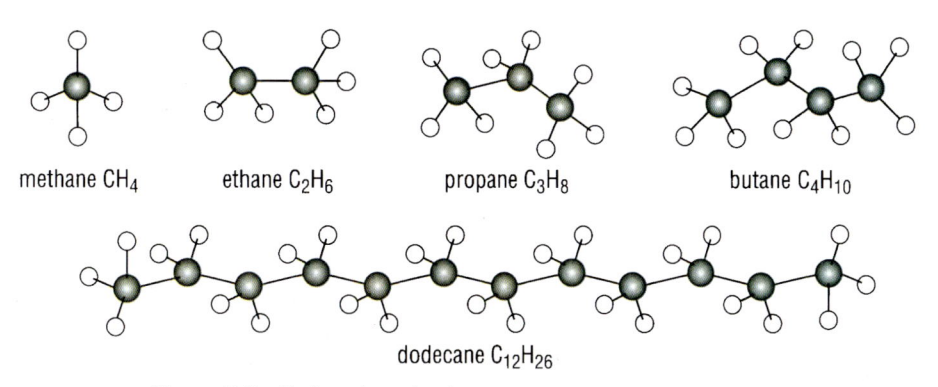

methane CH_4 ethane C_2H_6 propane C_3H_8 butane C_4H_{10}

dodecane $C_{12}H_{26}$

Figure 8.9 Hydrocarbon structures.

The different hydrocarbons have different boiling points and fractional distillation is used to separate them.

Fractional distillation of oil

1 Heating the oil
The oil is heated to about 450 °C and most of it turns into a vapour. This is introduced into the fractional distillation column which forms a tall tower, as Figure 8.10 shows.

2 Separating the hydrocarbons
The bottom of the tower is kept at 360 °C and the top of the tower is kept at 40 °C. The hot oil vapour

Figure 8.10 Distillation towers.

105

COMPOUNDS AND MIXTURES

is introduced into the tower below the mid-way point. Inside the tower are tiers of trays (see Figure 8.11). There are tubes called risers passing through each tray. Above each riser is a bubble cap.

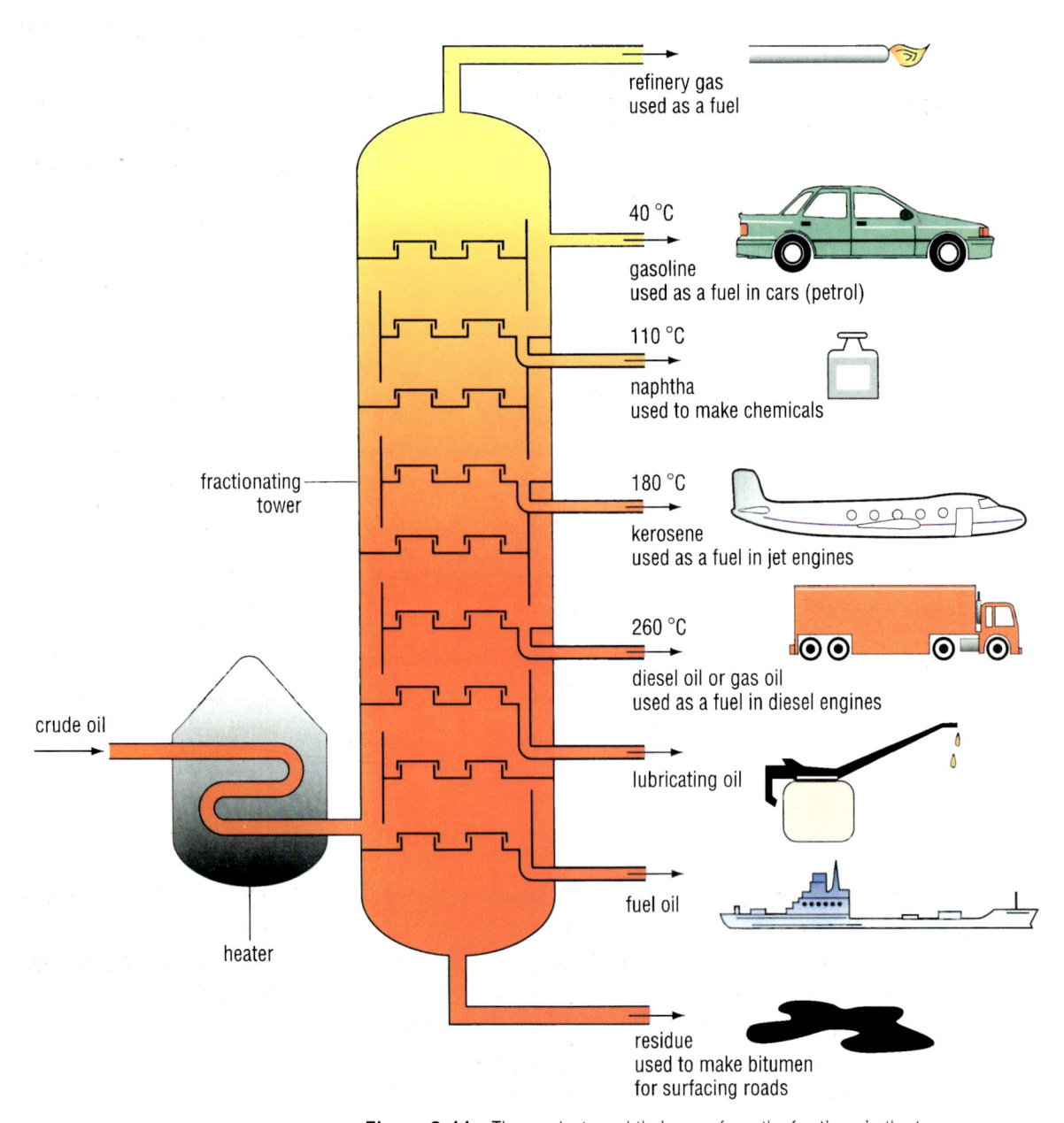

Figure 8.11 The products and their uses from the fractions in the tower.

When the vapour meets a tray, some of it condenses and forms a liquid. When the tray is full, some of the liquid spills down the overflow into the tray below. The hydrocarbons in the vapour that did not condense pass upwards through the risers and out under the bubble cap onto the tray above. This tray is slightly cooler than

106

COMPOUNDS AND MIXTURES

the one below and some of the hydrocarbons condense and form a liquid. When hydrocarbons condense they give out heat energy into the liquid they enter. This heat causes other hydrocarbons in the mixture, with lower boiling points, to evaporate and rise into the tray above.

The fractional distillation of oil is a continuous process and a fully operational tower has liquids in every tray. Gases bubble through them from the trays below while liquids containing hydrocarbons with the longest molecules move downwards through the overflow pipes. There are collection pipes at different heights up the tower. They collect different fractions of the oil. Each fraction is a mixture of hydrocarbons with similar boiling points and they are used to make up a range of products, as Figure 8.11 shows.

Table 8.1 The boiling points and range of number of carbon atoms in different oil fractions (simplified).

Fraction of oil	Boiling points of liquids °C	Number of carbon atoms
A	180	9–16
B	40	4–12
C	260	15–19
D	below 40	1–4
E	110	7–14

8 Table 8.1 shows the boiling points and range of number of carbon atoms in the molecules of five fractions of oil.
   a) Arrange the letters of the fractions in order, starting with the one that would be drawn from the top of the distillation tower.
   b) Look at Figure 8.11 to identify the fractions and write down a use for each one.
   c) What is the relationship between the boiling points of the fractions and the lengths of the hydrocarbons they contain?

9 How do gases move up the tower?

10 How do liquids move down the tower?

11 Which fractions are used for transport?

COMPOUNDS AND MIXTURES

♦ SUMMARY ♦

- Elements in a mixture keep their own properties (*see page 99*).
- Elements in a compound lose their own properties (*see page 99*).
- In a synthesis reaction two or more substances join together to make a compound (*see page 100*).
- The freezing point and the boiling point are two important physical properties of solutions (*see page 100*).
- A mixture of an insoluble and a soluble substance can be separated by filtration and evaporation (*see page 101*).
- A mixture of solutions can sometimes be separated by precipitation (*see page 101*).
- Electrolysis is the decomposition of an electrolyte using electricity (*see page 102*).
- Liquids with different boiling points can be separated by fractional distillation (*see page 104*).
- The hydrocarbons in oil can be separated by fractional distillation (*see page 105*).

End of chapter question

Imagine that you have to prepare a presentation on separating mixtures and compounds. What would you say about mixtures and their separation? What would you say about separating elements from a compound? Suggest practical demonstrations that you or your teacher could make during the presentation.

9 Metals and non-metals

The elements can be divided into metals and non-metals according to their properties. The way in which elements vary in a particular property can be found by comparing them and putting them in order.

Comparing hardness

You may compare the hardness of some elements such as copper, iron, carbon, sulphur or aluminium by scratching each one on the others. By doing this you will find that some elements scratch others and some do not. If an element scratches another element, it means that the element is harder than the one being scratched. If an element fails to scratch another one, it is softer than the element being scratched.

Comparing density

The density of a substance is a measure of the amount of matter in a certain volume of it. In the laboratory density is measured as grams per cubic centimetre – g/cm^3. Table 9.1 shows the densities of some metals and non-metals.

Table 9.1 Densities.

Metal	Density g/cm³
Mercury	13.6
Sodium	0.97
Copper	8.9
Zinc	7.1
Potassium	0.86
Silver	10.4
Calcium	1.55
Aluminium	2.7
Magnesium	1.74
Iron	7.8
Non-metal	
Oxygen	1.4
Hydrogen	0.09
Carbon (as graphite)	2.25
Silicon	2.49
Bromine	3.9
Iodine	4.9

Comparing melting point

Table 9.2 shows some melting points of metals and non-metals.

Table 9.2 Melting points.

Metal	Melting point °C
Sodium	98
Zinc	419
Potassium	63
Iron	1539
Copper	1083
Silver	962
Magnesium	650
Gold	1064
Calcium	850
Mercury	−39
Non-metal	
Oxygen	−219
Nitrogen	−210
Carbon	3650
Iodine	114
Hydrogen	−259
Bromine	−7
Silicon	1410

Exceptional metals

If you have tried questions 1 to 4 you will have found out that there are exceptions to the general conclusions you can make about the properties of metals and non-metals. Here are a few more exceptions related to other properties. Most metals are hard but sodium is so soft that it can be cut with a knife. Most metals are not magnetic but iron, nickel and cobalt are. Most metals are silvery white or silvery grey or blue. Gold and copper are exceptional

1. Arrange the metals and non-metals in Table 9.1 in order according to their densities. What general conclusions can you draw from your list?
2. What exceptions can you find to your general conclusions?
3. Arrange the metals and non-metals in Table 9.2 in order according to their melting points. What general conclusions can you draw from your list?
4. What exceptions can you find to your general conclusions?

METALS AND NON-METALS

because they are coloured. Gold is a bright yellow and copper is brown. All metals are solid at room temperature except one – mercury – which is a liquid.

Table 9.3 shows the physical properties of metallic and non-metallic elements.

Table 9.3 Physical properties of metals and non-metals.

Property	Metal	Non-metal
state at room temperature	solid	solid, liquid or gas
density	high	low
surface	shiny	dull
melting point	generally high	generally low
boiling point	generally high	generally low
effect of hammering	shaped without breaking	breaks easily
effect of tapping	a ringing sound	no ringing sound
strength	high	generally very weak
magnetic	a few examples	no examples
conduction of heat	good	poor
conduction of electricity	good	poor

Extracting and using metals

Metals differ from each other in the way they react with other elements and compounds. Some metals, such as gold and silver, are very unreactive and can be found in their metallic form in the Earth's crust. Elements which are found on their own in this way are called native elements. More reactive metals are found combined with other elements. A rock which is rich in a metal compound is called an ore.

Different methods of extracting metals are used. They depend on the reactivity of the metal. For example, copper is quite unreactive and so can be extracted from its ore by roasting the ore in a furnace. Iron, which is more reactive, must be heated strongly in a blast furnace. This provides more heat energy which is needed for the reactions to release iron from its ore.

In the sections that follow, the metals are arranged in order of their reactivity. The order starts with the least reactive.

METALS AND NON-METALS

Alloys

In the following descriptions of metals the word alloy will appear. Most metals in their pure form tend to be weak and soft. They are strengthened by mixing them with one or more other elements, usually different metals. These mixtures are called alloys. They are made by melting the metals, mixing them together and then allowing them to cool. The properties of an alloy can be changed by altering the proportions of the metals that are mixed together.

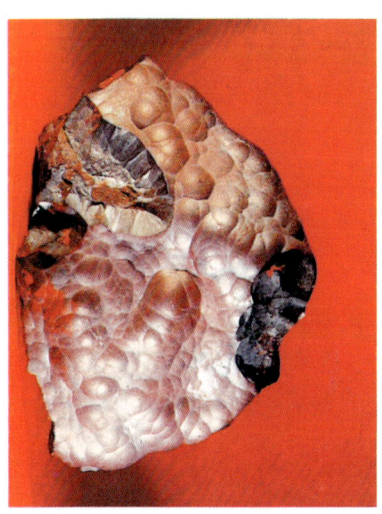

Silver crystals

Iron ore (haematite)

Figure 9.1 A native metal (silver crystals) and a metal ore (haematite).

Silver

Silver may be found on its own as a metal. Many pieces can occur together in veins (cracks in rocks) or they may be more spread out in the ores of other metals. A black mineral called silver glance, which is formed from silver sulphide, may also be found with the silver metal. Silver that has already been used is recycled. Coins containing silver and industrial wastes, particularly from the photographic industry, are sources of recycled silver.

Extraction

Most silver extracted today is collected during the purification of copper, zinc and lead ores.

Properties and uses of silver

Silver has the highest reflectivity of light of any metal. This means that its surface reflects more of the light shining onto it than the surface of other metals.

111

METALS AND NON-METALS

This property makes it particularly attractive for use in jewellery, cutlery and ornaments. It is a soft metal in its pure form and is hardened by alloying it with copper to make sterling silver.

Electroplated nickel silver or EPNS is made from nickel silver (an alloy of copper and nickel) that is covered in a thin coating of silver.

5 Why do you think sterling silver is better for cutlery than pure silver?

Figure 9.2 Silver objects.

Copper

The ore from which copper is extracted is called chalcopyrite or copper pyrites. It contains copper, iron and sulphur, is brass yellow, and is found in igneous and metamorphic rocks.

Extraction

The ore is concentrated in a flotation cell (see page 67) and the copper, iron and sulphur are separated by roasting the ore in a furnace. The copper that is removed from the furnace still contains impurities. They are removed by making the copper into large slabs and hanging them in an electrical cell (see Figure 9.3). Each slab is an anode. As the electricity passes through the cell, the copper at the anode dissolves in the electrolyte and comes out of solution again on the cathode where it forms pure copper.

Gold and silver impurities in the metal fall to the bottom of the cell below the anode and form the anode sludge. They are removed and separated.

6 What are the three processes used in the extraction of copper?
7 What other metals are there in copper ore?

METALS AND NON-METALS

Figure 9.3 Making pure copper by electrolysis.

Properties and uses

Copper is a soft metal that can be easily shaped. It does not react with water so it is used to make water pipes, though large water pipes are often made of plastic which is cheaper. Copper conducts heat well and is used in the bases of some kinds of kitchen pan. Copper's softness also allows it to be pulled out into a wire and as it also conducts electricity well and corrodes very slowly, the wire can be used to conduct electricity inside buildings.

Copper is alloyed with tin to make bronze. This alloy was first made and used 5000 years ago and its name is used to describe a period of history in which a great many bronze implements were used – the Bronze Age. Bronze is a particularly sonorous metal (it makes a ringing sound) and it is used to make bells and cymbals because of the clear ringing sound it produces when it is struck.

Brass is an alloy of copper and zinc. It is strong, corrosion-resistant and is used to make the pins in electrical plugs. It is also a shiny metal and is used to make ornaments.

8 What properties of copper make it useful for electrical wiring in the home?

9 Why is brass better for use in plug pins than copper?

Figure 9.4 A 16th Century Benin bronze.

113

METALS AND NON-METALS

Iron

Iron frequently combines with oxygen to form iron oxide. In the Earth's crust this compound frequently forms the mineral haematite (see Figure 9.1 page 111). Sedimentary rocks which contain large amounts of haematite are important iron ores.

Extraction

Iron is separated from the oxygen in iron oxide by a reduction process. This takes place in a blast furnace (see Figure 9.5). The iron ore is mixed with coke and limestone and tipped into the top of the blast furnace. Hot air is blown into the blast furnace through pipes close to the base of the furnace.

The hot air causes the coke to ignite and burn strongly, raising the temperature to as high as 2000 °C. The carbon in the burning coke reacts with oxygen in the hot air to form carbon dioxide. This gas rises through the hot coke higher up the furnace and reacts with the carbon in it to form carbon monoxide. As the carbon

10 Write word equations for the reactions which take place inside the blast furnace. Make a simple sketch of the furnace and indicate on it where each of the reactions takes place.

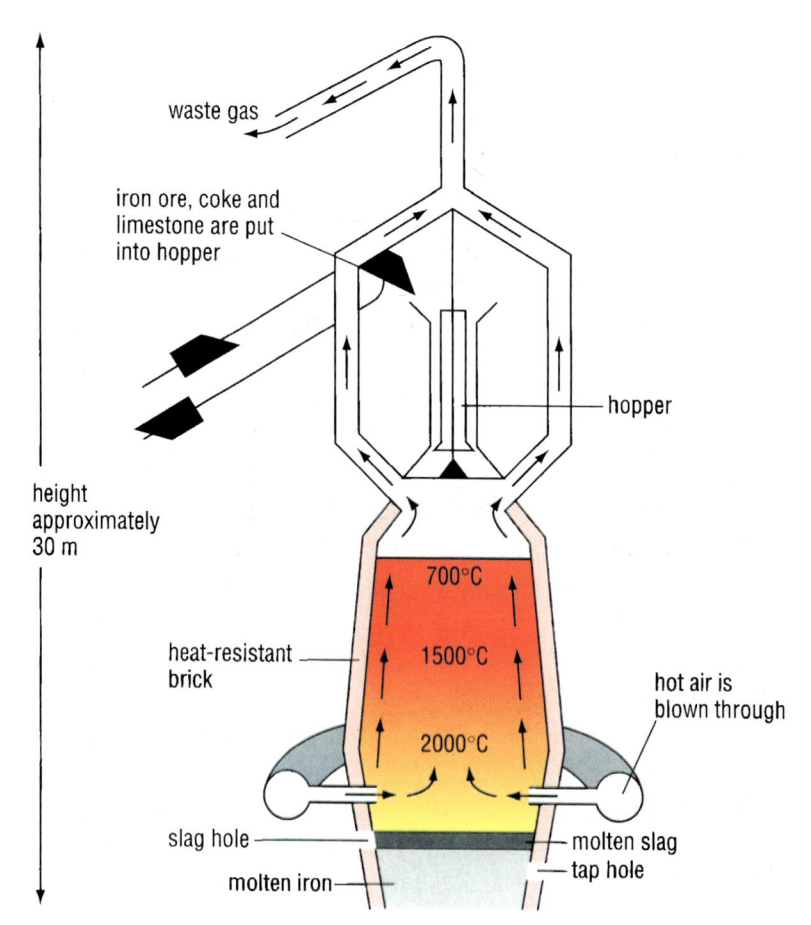

Figure 9.5 A blast furnace.

METALS AND NON-METALS

11 Why is limestone added to the ore and the coke?
12 How many tonnes of iron are produced in a blast furnace, if it produces 10 000 tonnes per day non-stop for 10 years?
13 How does using the gases from the top of the blast furnace to heat the incoming air make the iron cheaper to produce?
14 How do you think the blast furnace got its name?

monoxide rises in the furnace it reacts with the iron oxide in the ore to produce carbon dioxide and iron. The metal is in a solid form which has tiny holes in it.

As the iron sinks down the blast furnace it gets hotter and melts. The limestone also sinks with the iron and the heat causes the calcium carbonate from which it is made to break down into calcium oxide and carbon dioxide. The rocky substance in the iron ore is silicon oxide, which does not melt at the high temperatures in the furnace. However the calcium oxide combines with the silicon oxide to make calcium silicate, which is known as slag. This substance melts in the high temperatures of the blast furnace and flows out with the molten iron. The slag floats on the molten iron and is easily separated from it.

Pig iron

The iron which is drawn out of the bottom of the blast furnace is run down a channel which has a series of moulds branching from it on one side. This arrangement of channel and moulds is similar to the way piglets lie when they feed from their mother and the metal that flows into the channel and moulds is called pig iron.

Early iron workers

Very rarely, iron occurs as a native metal. Some metal iron is formed where hot volcanic rock meets a seam of coal. The heat allows a chemical reaction to take place between the iron compounds in the rock and carbon in the coal. Many meteorites that strike the Earth are made of iron. It is possible that early people knew of metal iron but as the pieces were so rare they did not begin to make iron products on a large scale.

A chance heating of an iron ore in a charcoal fire most probably led to the discovery of the extraction of iron and the development of metal products. Before this discovery was made, bronze was the most common metal in everyday use. It is an alloy of copper and tin, but it is quite weak. Soldiers using bronze swords in battle had to stop occasionally to straighten them!

Iron is a stronger metal than bronze and soon replaced it as the most common metal for everyday uses.

The Hittites were the first people to produce iron in large amounts about 3500 years ago. They were a people that lived in the land now called Turkey.

Iron needed a higher temperature than copper or tin for extraction from its ore. The extraction was achieved by heating the ore with charcoal and using bellows to supply more air to the furnace.

(continued)

METALS AND NON-METALS

The iron made in these early furnaces was wrought iron. This is a soft form of iron. The Hittites discovered how to give the wrought iron a coating of harder metal by allowing the iron surface to combine with some of the carbon in the charcoal and form steel.

In Northern India, the iron workers developed a process which prevented wrought iron from rusting. In 400 BC they made a pillar of wrought iron 8 metres high and 6 tonnes in weight. This pillar is still standing today and does not have any rust (see Figure A). Nobody knows how these early iron workers made such a metal.

1 How do you think the discovery of iron affected warfare between countries?
2 What substance in the air takes part in a chemical reaction in the furnace?
3 How did the use of bellows help in the extraction of iron?
4 Steel is an alloy. From what substances is it made?
5 How did the discovery made by iron workers in Northern India differ from that made by the Hittites?

Figure A This wrought iron pillar was built in 400 BC and is still standing.

Properties and uses of cast iron

When pig iron is re-melted it can be poured or cast into more complicated moulds and is known as cast iron. As the metal cools it expands a little and fills every part of the mould. This makes it suitable for use in complicated moulds like those used to make car engine blocks (see Figure 9.6, opposite). Cast iron is also strong and is used for manhole covers in the street since the metal can support the weight of traffic running over it. However, cast iron is brittle. This means that the metal breaks easily if it is bent, so it cannot be shaped by bending after it has cooled and set.

METALS AND NON-METALS

> **For discussion**
>
> Most iron is converted into steel. This is done by blasting oxygen into the hot metal to remove carbon.
>
> A range of steels can be produced depending on the amount of carbon removed from the iron. Each type of steel has its own special properties. Do we live in a 'steel world'? Look around you and use sources like the internet to form an opinion. Do others agree or disagree with you?

Figure 9.6 A cast iron engine block.

Extracting non-metals

The gases of the air are non-metals. Figure 9.7 shows the proportions of the gases in air. They are extracted by the fractional distillation of liquid air (see pages 104 and 105).

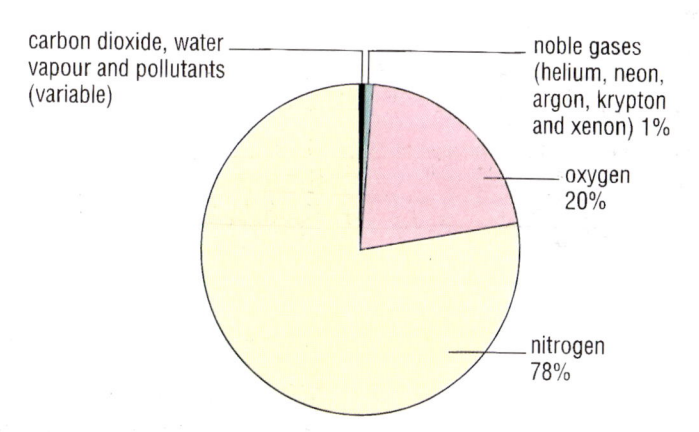

Figure 9.7 The composition of the air near the Earth's surface.

The non-metals carbon and sulphur are extracted from the ground and have useful properties.

Carbon

Most of the carbon in the Earth's crust is combined with other elements to make compounds called carbonates. Carbon also exists in pure forms which have different structures. These different forms are called allotropes. The three allotropes of carbon are graphite, diamond and buckminsterfullerene. Charcoal is an impure form of carbon which is a useful material.

METALS AND NON-METALS

Graphite

Graphite is found in some igneous and metamorphic rocks. It is also made artificially by using electricity to heat coke for several hours to a temperature of about 2500 °C. This way of producing graphite artificially is called the Acheson process.

The way the atoms of carbon join together to form graphite is shown in Figure 9.8. The carbon atoms form layers. In each layer the carbon atoms are held strongly together to form hexagonal structures. The layers are held together by only weak forces.

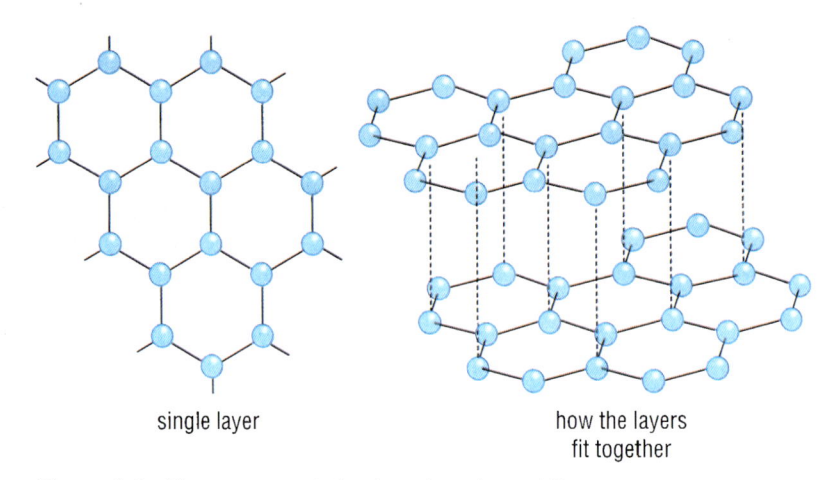

single layer   how the layers fit together

Figure 9.8 The arrangement of carbon atoms in graphite.

Properties and uses

Graphite is used to make the 'lead' in pencils. When a pencil point is drawn across the surface of paper the forces pulling on the graphite are stronger than the forces holding the layers together and the layers slide off each other onto the paper to make a pencil mark. Graphite in pencils is mixed with clay to give the 'lead' strength. The slipperiness of graphite also makes it a good lubricant to reduce the friction between the moving parts of machinery.

Diamond

In the Earth's crust in Kimberley in South Africa there are pipes of rock. They have formed from molten rock rising up through the crust. The rock is called kimberlite and in some of the rock diamonds have formed. In other places diamonds have been found in the gravel in rivers and on beaches. They have been released from rock when the rock weathered.

15 Do you think a hard pencil has more clay or less clay than a soft pencil?

METALS AND NON-METALS

Diamonds are made artificially by mixing carbon with nickel and squashing the mixture to 50 000 times atmospheric pressure and raising the temperature to 1500 °C. After a few minutes very small diamonds are formed which can be used for industrial purposes.

The arrangement of carbon atoms in diamond is shown in Figure 9.9. The forces acting on each carbon atom are the same strength in each direction. This makes a very hard substance.

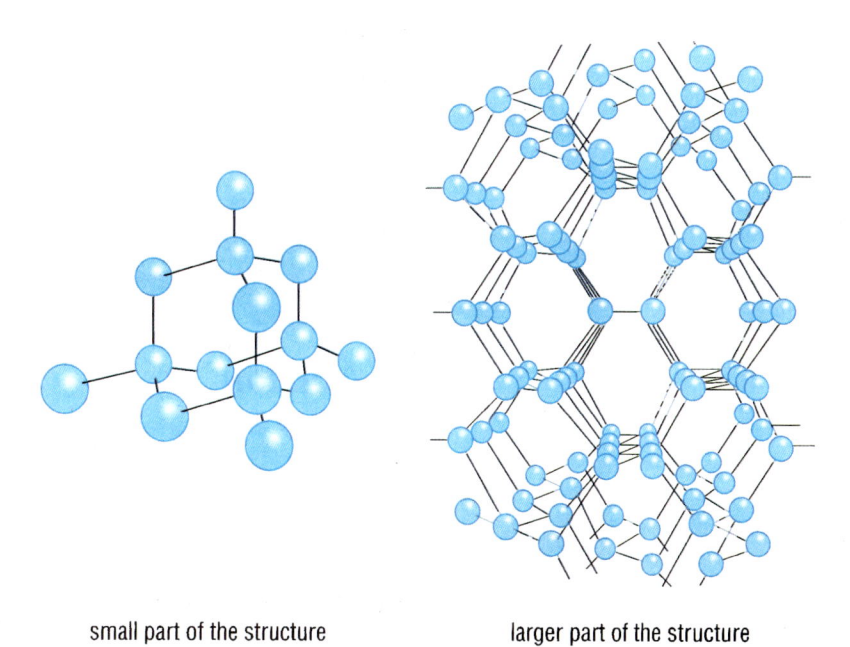

small part of the structure   larger part of the structure

Figure 9.9 Arrangement of carbon atoms in diamond.

Properties and uses

Diamonds are used to drill through rock in search of oil and to cut through concrete, glass and metal. Most diamonds are dark and opaque but some are transparent. These diamonds are carefully cut to reflect the light and are highly prized for the way they sparkle. They are used to make jewellery.

Charcoal

Wood charcoal is made by heating wood in the absence of air. In the past it was an important fuel and was used 3500 years ago to reduce iron oxide in iron ore to obtain iron metal. Charcoal absorbs smoke and this property makes it a suitable fuel for cooking food in barbecues.

METALS AND NON-METALS

16 Compare graphite and diamond.

17 Carbon has a wide range of uses. Assess this statement and give examples.

Charcoal absorbs poisonous gases and is used in gas masks. It is used in aquarium filters to clean the water in fish tanks and in oven hoods to remove the smells of cooking in kitchens.

Coke

Coke is made by heating coal without air. It is used to reduce the oxides of metals in the production of metals such as iron, zinc and lead.

Sulphur

Sulphur is an element that escapes as a gas from the vent of a volcano. The solid sulphur forms a crust on rocks in the volcano vent. Large amounts of sulphur are also found in sedimentary rocks such as limestone. The sulphur moves into the sedimentary rocks from volcanic regions by melting and flowing through the crust over a long period of time.

Extraction

Sulphur is sometimes extracted from natural gas but most of it is extracted from limestone rocks in the USA. The rocks are at least 100 metres below the surface and are covered in sand. Mining is not possible because mine shafts cannot be safely constructed in sand as the walls easily collapse. In addition, if the limestone rocks could be reached the smell of the sulphur would be unbearable and sulphur dioxide, a poisonous gas, would be produced as the rocks heated up during drilling. In 1894, Herman Frasch (1851–1914) invented a sulphur pump to extract sulphur from the rocks (see Figure 9.10).

The pump works because sulphur has a melting point of 115 °C and enough heat can be supplied to the rock by super-heated water to melt the sulphur and pump it out. Super-heated water is made by heating water under pressure until its temperature reaches 170 °C.

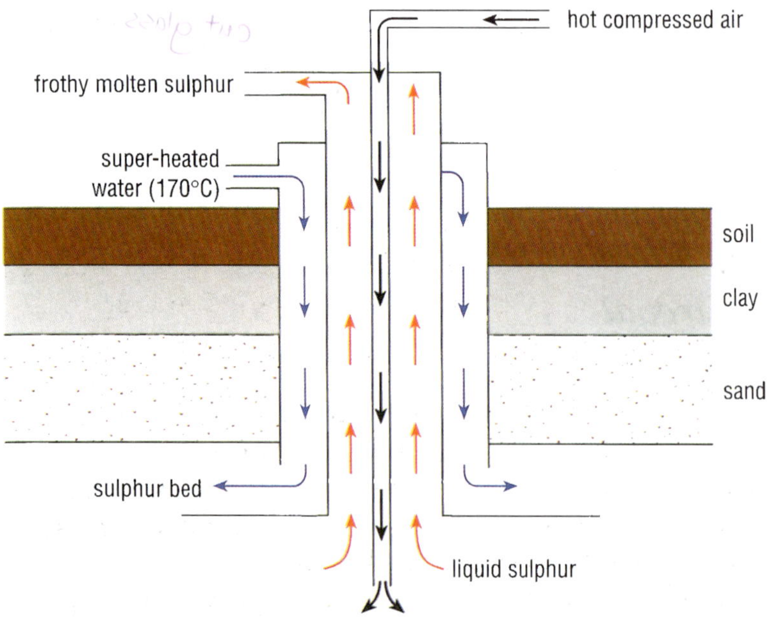

Figure 9.10 The Frasch sulphur pump.

METALS AND NON-METALS

There are three pipes arranged concentrically (one around the next) which connect the surface of the ground to the beds of sulphur in the rock. Super-heated water is pumped down the outer pipe and into the sulphur bed. Its heat transfers to the sulphur and melts it. Compressed air is pumped down the central pipe. Its pressure is 15 times that of the air and it pushes the mixture of molten sulphur and hot water up the middle pipe. The sulphur and water emerge from the pipe as a froth. This is poured into a vat where the water and sulphur cool. The water is drained off the solid sulphur. The sulphur is almost pure and can be used straight away for industrial purposes.

18 Draw a cross-section of the pipes used in the sulphur pump. State what each contains and the direction of its movement.

Figure 9.11 Sticks of sulphur extracted by the Frasch process.

Properties

Sulphur is a yellow solid at room temperature. It is brittle and does not dissolve in water. It burns in the air to form sulphur dioxide. This gas is used in the manufacture of sulphuric acid (see page 88).

Sulphur reacts with rubber when the two are heated together. It binds the long molecules of rubber together and stops them sliding past each other. Rubber treated with sulphur is called vulcanised rubber. It is much harder than natural rubber and tough enough for use in tyres on all kinds of cars and trucks.

Sulphur is poisonous to fungi and powdered sulphur is used as a fungicide.

■ METALS AND NON-METALS

Chemicals and the land environment

The major chemical pollutants on land are pesticides which can affect human health, and radioactive chemicals accidentally released from nuclear power plants which can cause cancer to develop. DDT is a pesticide that causes serious long-term problems and it is now banned in many countries. However, it is still used in some countries where no laws exist to restrict its use. They continue to use it because it is very effective in controlling the mosquitoes that spread malaria.

The discarded products of manufacturing industries produce a pollution problem in every country. The tips in which the waste is stored take up space.

Today, many tips are carefully filled so that when they are full they can be covered with soil and new habitats established on top of them. While the rubbish is settling and decomposing on the tip some of it gives off methane gas. This can be collected by a system of pipes and used as a fuel.

Figure 9.12 A tip with a methane 'breather'.

Most raw materials have to be taken out of the ground. In some cases mine shafts are sunk into the ground and the material is removed with little damage to the surrounding habitat. Lead, zinc and some copper and coal are mined in this way.

In open cast mining, the land surface is removed to extract the raw material (see Figure 9.13). Aluminium and some coal and copper are removed like this. It causes complete habitat destruction. If this occurs in rainforest areas the forest may not be able to grow back again when the mining operation is over because the thin layer of soil on which the forest grew may have been completely washed away.

19 How do the methods of extracting raw materials affect plants and animals that live in the same area?

METALS AND NON-METALS

Figure 9.13 Open cast mining in a rainforest in South America.

20 Many natural forests have a mixture of many different species of tree. They are of different ages and are irregularly spaced out. Many planted forests have very few tree species. The trees are the same age and are regularly spaced out.
  a) In what ways are the planted forests different from the natural forests?
  b) Do you think the two forests will support the same wildlife? Explain your answer.

Figure 9.14 A coniferous forest being replanted.

Renewable and non-renewable materials

Raw materials can be divided into two groups – renewable materials and non-renewable materials. Wood is an example of a renewable material. As trees are cut down to provide the raw material for wood products, young trees are planted to replace them. Iron is an example of a non-renewable material. There is a certain amount of it in the Earth's crust which is not replaced after it is used up. As the iron ore is mined the supply left in the ground is reduced. In time there could be none left to use.

As the human population increases, the demand for raw materials also increases. Although renewable materials can be replaced, the extra demand means that extra space has to be found for the material to be re-formed. This can result in habitat destruction. An example of this is where moorlands are planted with forests of fast-growing trees to be used in manufacturing.

As non-renewable raw materials cannot be replaced, studies have been made to find out how much of each material is left on the Earth. For the purpose of the study the raw materials are divided into three groups. These are the stocks, the reserves and the resources. The stocks are the materials which are already mined and

METALS AND NON-METALS

21 If the world stock of a material is 10 000 000 tonnes and it is used at a rate of 250 000 tonnes a year:
  a) how long will the stocks last,
  b) when will the stocks run out?

stored ready for use. The reserves are materials still in the ground that can be mined economically (they are not too expensive to extract). The resources cannot be mined economically – they are too expensive to mine at present. (Note that another use of the word resource in science is to mean a supply that is readily available.) Once stocks are used up, reserves will be mined and converted into stocks. Resources may then become reserves and a material comes closer to being used up. This process can be slowed down so the material is conserved by recycling.

Recycling

In many parts of the world people recycle materials by simply using them in another way. For example, when car tyres are worn out they can be cut up and used as the soles of shoes. Many people prefer to throw away items when they are no longer useful. A newspaper may be read for a day, a bottle of lemonade may last three days, an item of clothing may last a year and a car may last 15 years. If the products are thrown away when their use is over, the materials in them just stay in the ground in a tip. They take up space and have to be replaced by extracting more raw materials and using large amounts of energy in the manufacturing processes. Recycling the materials saves space, raw materials and energy.

Figure 9.15 This person is wearing shoes made from old car tyres.

METALS AND NON-METALS

22 Imagine that a new product has been invented that uses the material in question 21. An extra 30 000 tonnes a year of the material is extracted for this product.
  a) How long will the world stocks now last?
  b) When will the stocks now run out?

23 Imagine a recycling programme has been set up in which 200 000 tonnes of the material in question 21 could be recycled each year.
  a) How long would the stocks now last using:
    i) 250 000 tonnes a year,
    ii) 280 000 tonnes a year?
  b) What effect does the recycling programme have on the reserves of the material?

24 If 1000 million tonnes of bauxite are mined every year and it is estimated that stocks will last until about 2240, how much bauxite is on the Earth?

25 What are the benefits of recycling?

Paper is made from smashing wood into a pulp of tiny fibres and then binding them together in a thin sheet. When paper is recycled it is made into a pulp of fibres again, without having to use energy and chemicals to break down the wood.

Glass is made from sand, limestone and soda, and a large amount of heat energy is required. Less energy is needed to melt recycled glass and make it ready for use again. The recycled glass is mixed with the raw ingredients as new glass products are made.

Large amounts of energy are needed for the extraction of metals such as iron (see page 114) and aluminium. Less energy is needed to melt them down than to extract new metals from their ores. By recycling metals, less fuel is used and stocks of the ores are conserved.

Methods of separation

Materials for recycling can be separated by people and taken to recycling centres (see Figure 9.16) or they can be separated after the collection of refuse. The magnetic separator is used to separate iron and steel from other materials.

In industry, products which are wastes in one process can be collected and used elsewhere. For example, in the purification of copper the metals silver and gold are produced (see page 112). These metals are not discarded but sold to people such as jewellery manufacturers who can use them.

Figure 9.16 Recycling centre.

Using energy efficiently

Some of the reactions which take place in the chemical industry produce heat energy. This is not released but used in other parts of the chemical plant. For example,

METALS AND NON-METALS

the heat produced when sulphur and oxygen combine in a combustion reaction is used to melt the solid sulphur at the beginning of the process to manufacture sulphuric acid.

Materials and energy

The processing of all materials needs energy and this is provided mainly by the fossil fuels – coal, oil and natural gas. These are non-renewable raw materials and while stocks of coal may only last until the year 2300, stocks of oil and natural gas are predicted to be used up in your lifetime, if used at the present rate. When materials are recycled there is a reduction in the amount of energy used to make the new products. Although some energy is used in the recycling process, it is usually less than the energy used in extraction.

Using materials in the future

Increasing amounts of many materials are being recycled and new ways are being found to save energy in chemical processing to meet the demands of the human population, today and in the future. New materials are made every year through investigations into the way different chemicals react together. From these discoveries, materials are selected which can perform a task more efficiently than an existing material and require smaller amounts of raw materials and energy. In the long term, there are plans to set up mines on the Moon to extract minerals and to process them, to make materials in space for use on Earth and in further space exploration.

26 When oil and natural gas supplies are used up, do you think the stocks of coal will still be expected to last until 2300? Explain your answer.

27 How does the recycling of materials affect the stocks of fossil fuels? Explain your answer.

Figure 9.17 Impression of a manned lunar base.

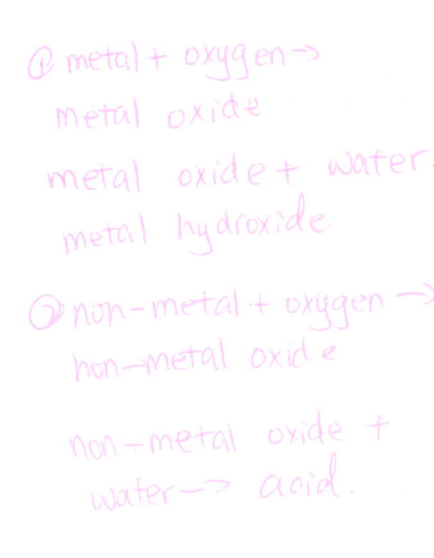

Chemical properties of metals and non-metals

Metals reacting with non-metals

Some metals and non-metals react together to produce salts. These reactions are examples of synthesis reactions (see page 100). For example, if a burning piece of sodium is placed in a jar of chlorine gas in a fume cupboard the two elements combine to make a white solid. The word equation for this reaction is:

sodium + chlorine → sodium chloride

If zinc or copper is heated with sulphur the metal sulphide is formed. The word equations for these reactions are:

zinc + sulphur → zinc sulphide

copper + sulphur → copper sulphide

Oxygen is a non-metal and reacts with many metals and non-metals to form oxides.

Reaction with oxygen

Metals

If a metal takes part in a chemical reaction with oxygen, a metal oxide is formed. A metal oxide is a base (see page 22) and forms a salt and water when it takes part in a chemical reaction with an acid. A few metal oxides are soluble in water. They are called alkalis. Calcium oxide is a soluble base (an alkali). This is the reaction that occurs between calcium oxide and water:

calcium oxide + water → calcium hydroxide

Non-metals

If a non-metal takes part in a chemical reaction with oxygen it also forms an oxide. Most oxides of non-metals are soluble. When they dissolve in water they form acids. Sulphur is a non-metallic element with a yellow crystalline form. If it is heated in air it burns and combines with oxygen to form sulphur dioxide, which is soluble in water. This reaction occurs between sulphur dioxide and water:

sulphur dioxide + water → sulphurous acid

METALS AND NON-METALS

28 How may the reaction with oxygen be used to distinguish a metal from a non-metal?

29 Use the information in this section to decide whether:
   a) carbon, and
   b) magnesium
   is a metal or a non-metal. Explain your answer.

When carbon powder is heated in air it glows red. If it is plunged into a gas jar of oxygen it becomes bright red. Carbon combines with oxygen to form carbon dioxide, which dissolves in water to form an acidic solution with a pH of 5.

Magnesium ribbon easily catches fire if it is held in a Bunsen burner flame, and burns with a brilliant white light if plunged into a gas jar of oxygen. Magnesium oxide (a white powder) is produced, which dissolves in water to make an alkaline solution with a pH of 8.

Reaction with water

Metals

Here are some descriptions of the reactions that take place between water and metals. In the study of these reactions the metals were first tested with cold water. If there was no reaction, the test was repeated with steam (see Figure 9.18).

Calcium sinks in cold water and bubbles of hydrogen form on its surface, slowly at first. The bubbles then increase in number quickly and the water becomes cloudy as calcium hydroxide forms. The bubbles of gas can be collected by placing a test-tube filled with water over the fizzing metal. The gas pushes the water out of the test-tube. If the tube, now filled with gas, is quickly raised out of the water and a lighted splint held beneath

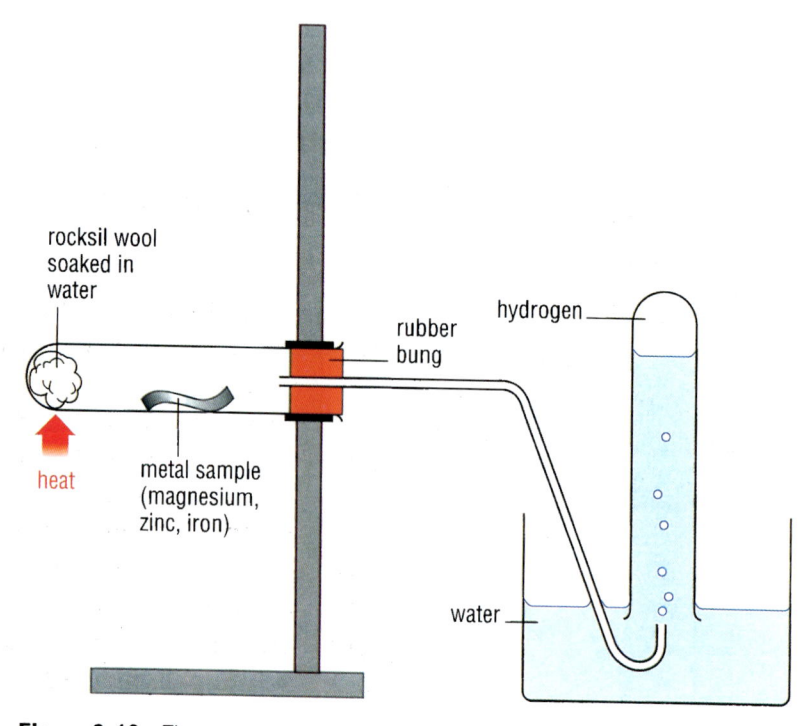

Figure 9.18 The apparatus used to test the action of steam on a metal.

METALS AND NON-METALS

30 Arrange the metals in this section in order of their reactivity with water.

31 Which metals would not be put into the apparatus in Figure 9.18 to see if they reacted with steam?

32 Which metals are less dense than water?

33 Water is a compound of hydrogen and oxygen which could be called hydrogen oxide. When hydrogen is released as a metal reacts with steam, what do you think is the other product of the reaction?

34 In the home, copper is used for the hot water tank and steel (a modified form of iron) is used to make the cold water tank. Why can steel not be used to make the hot water tank?

35 Arrange the metals in this section in order of their reactivity with hydrochloric acid.

36 Why was a concentrated solution used if there was no reaction with a dilute solution?

37 If a metal which had reacted very slowly with a dilute acid was tested with a concentrated one, what would you predict would happen?

38 Construct a general word equation for the reaction between a metal and hydrochloric acid. (Instead of using the name of a metal, just use the word 'metal'.)

its mouth, a popping sound is heard. The hydrogen in the tube combines with oxygen in the air and this explosive reaction makes the popping sound.

Copper sinks in cold water and does not react with it. Neither does it react with steam.

Sodium floats on the surface of water and a fizzing sound is heard as bubbles of hydrogen gas are quickly produced around it. The production of the gas may push the metal across the water surface and against the side of the container, where the metal bursts into flame. A clear solution of sodium hydroxide forms.

Iron sinks in water and no bubbles of hydrogen form. When the metal is heated in steam, hydrogen is produced slowly.

Magnesium sinks in water. Bubbles of hydrogen are produced only very slowly and a solution of magnesium hydroxide is formed. When the metal is heated in steam hydrogen is produced quickly.

Potassium floats on water and bursts into flames immediately. Hydrogen bubbles are rapidly produced around the metal. A clear solution of potassium hydroxide forms.

Non-metals

Carbon

The atoms of carbon are held very strongly together (see pages 118 and 119). This means that they do not enter into a chemical reaction with water and are insoluble in it.

Sulphur

Like carbon, sulphur does not take part in a chemical reaction with water and is insoluble in it.

Reaction with acid

Metals

On page 87 the action of hydrochloric acid on metals was described to show how metal chlorides can be made.

Here are some descriptions of the reactions that take place between different metals and hydrochloric acid. The metals were first tested with dilute hydrochloric acid. If a reaction did not take place, they were tested with concentrated hydrochloric acid.

Lead did not react with dilute hydrochloric acid but when tested with concentrated acid, bubbles of hydrogen gas were produced slowly and a solution of lead chloride was formed.

METALS AND NON-METALS

Zinc reacted quite slowly with dilute hydrochloric acid to produce bubbles of hydrogen and a solution of zinc chloride was formed.

Copper did not react with either dilute or concentrated hydrochloric acid.

Magnesium reacted quickly with dilute hydrochloric acid and formed bubbles of hydrogen and a solution of magnesium chloride (see Figure 9.19).

Figure 9.19 The reaction of magnesium with dilute hydrochloric acid produces hydrogen bubbles.

Iron reacted slowly with dilute hydrochloric acid to produce bubbles of hydrogen and a solution of iron chloride was formed.

Non-metals

Carbon

Carbon reacts with hot, concentrated nitric acid as the word equation describes:

carbon + nitric → water + nitrogen + carbon
acid   dioxide   dioxide

Nitrogen dioxide is a brown, poisonous gas. It condenses at 210 °C to form a yellow liquid and at −100 °C it forms colourless crystals.

Carbon reacts with hot sulphuric acid as the word equation shows:

carbon + sulphuric → water + sulphur + carbon
acid   dioxide   dioxide

Sulphur dioxide is a colourless, poisonous gas.

METALS AND NON-METALS

Sulphur

Sulphur reacts with hot, concentrated nitric acid as the word equation shows:

sulphur + nitric acid → sulphuric acid + nitrogen dioxide + water

Sulphur reacts with hot, concentrated sulphuric acid as the word equation shows:

sulphur + sulphuric acid → sulphur dioxide + water

◆ SUMMARY ◆

- Metals and non-metals have different physical properties (*see page 109*).
- A few metals are found uncombined with other elements in the Earth's crust, but most are present in compounds (*see page 110*).
- Metals can be mixed to form alloys (*see page 111*).
- Silver has properties that are useful for decorative purposes (*see page 111*).
- Copper is purified by electrolysis (*see page 112*).
- Iron is extracted from iron oxide by a reduction reaction in the blast furnace (*see page 114*).
- There are three pure forms of carbon. They are graphite, diamond and buckminsterfullerene (*see page 117*).
- Charcoal is a useful form of impure carbon (*see page 119*).
- Coke is used to reduce the oxides of metals during metal production (*see page 120*).
- Sulphur is extracted using super-heated water and has a wide range of uses (*see page 120*).
- Waste is stored in tips (*see page 122*).
- Some raw materials are extracted by open cast mining (*see page 122*).
- Some materials are renewable while others are non-renewable (*see page 123*).
- Recycling conserves stocks of raw materials including fuels (*see page 124*).
- New materials are being made all the time using more energy-efficient processes (*see page 125*).
- Some metals and non-metals react together to produce salts (*see page 127*).
- If a metal or non-metal reacts with oxygen an oxide is formed (*see page 127*).
- The reactions between metals and water vary in their speed (*see page 128*).
- Carbon and sulphur do not react with water (*see page 129*).
- The reactions between metals and acids vary in their speed (*see page 129*).
- Carbon and sulphur react with hot, concentrated nitric acid and sulphuric acid (*see page 130*).

METALS AND NON-METALS

End of chapter questions

1 Construct a table that summarises the information about the metals and non-metals in this chapter. Use headings such as source, extraction method and properties.

2 Using the information in this chapter, what policies would you suggest to the governments of all countries to improve the quality of the land environment and to ensure that future generations have sufficient resources to meet their needs?

10 Corrosion

Fast reactions

Very occasionally there may be a news story about an explosion in a coal mine or in a flour mill. Explosions in these places can be caused by the coal and flour themselves and not explosives such as dynamite. The explosion is due to the materials forming a dust in the air.

If a piece of coal is heated, it produces a flame and burns steadily in air. If coal dust is heated, it explodes. The reason for this difference in reaction is due to the surface area of coal in contact with the air. When a piece of coal is ground into dust it has a much larger surface area in contact with the oxygen in the air. This means that once the coal dust is hot it can react very quickly. The heat produced by this fast reaction causes the air to expand rapidly and push on everything around it with great force.

1 A teacher sets up a candle in a tin and places some cornflour powder next to it in the tin. She lights the candle and puts a lid on the tin and then blows into it through a tube to make the cornflour form a dust around the candle.
   a) What do you think will be heard?
   b) What do you think will be seen?
   c) Why have these changes taken place?

Figure 10.1 A coal mine where dust-reduction methods are used.

Slow reactions

Some foods contain fats and oils. If they are left out in the air for a few days they may start to smell unpleasant. This is due to a chemical reaction which has taken place between the oils and fats and the oxygen in the air. It results in the food becoming rancid and inedible.

A chemical reaction in which oxygen combines with a substance is called an oxidation reaction. Chemicals have been developed to slow down the oxidation of fats and oils in air. They are called antioxidants.

CORROSION

2 Have a look at the ingredients listed on the packaging of a range of prepared foods.
   a) How many of the foods have antioxidants added to prevent the food going rancid?
   b) What percentage of the foods you examined have antioxidants in them?

Many foods such as corn chips are further protected from oxidation by the gas in their unopened bags. It is nitrogen. As oxygen is not present in the bags oxidation cannot take place. Once the bags are open the antioxidants in the food prevent the food from going rancid.

INGREDIENTS: Sugar, Glucose-Fructose Syrup, Water, Lemon Juice (7%), Modified Maize Starch, Margarine, Eggs, Citric Acid, Acidity Regulator: Sodium Citrates, Lemon Oil (0.17%), Gelling Agent: Pectin, Antioxidant: Ascorbic Acid, Colour: Beta Carotene.

Figure 10.2 This food contains an antioxidant.

3 When a piece of iron rusts, would you expect it to weigh more or less than the original piece of iron? Explain your answer.

Rusting

Rusting is a slow reaction which takes place between iron and oxygen. It is an oxidation reaction. The word equation for the reaction is:

$$\text{iron} + \text{oxygen} \rightarrow \text{iron oxide}$$

When water vapour from the air condenses on iron or steel it forms a film on the surface of the metal. Oxygen dissolves in the water and reacts with the metal to form iron oxide. This forms brown flakes of rust which break off from the surface and expose more metal to the oxygen dissolved in the water. The iron or steel continues to produce rust until it has completely corroded.

Steel is used for making girders that support bridges and for making many parts of cars. If the steel is not protected it soon begins to rust. This weakens the metal. It makes bridges unsafe. It makes holes in car bodies and weakens the joints that hold the cars together, making them unsafe for use. Steel is a compound of iron and carbon. The iron in this compound rusts when exposed to damp air.

Figure 10.3 A rusted iron gate post.

CORROSION

4 On a winter's day a car is driven through slushy streets treated with salt (see page 100) and then parked in a warm garage.
   a) How will the rate of rusting compare with a car that has not been driven in the slush but left parked outside? Explain your answer.
   b) What advice would you give to the driver of the first car to slow down rusting?

5 Many tall buildings have a framework made of steel girders on which walls of brick and glass are built. If the steel was unprotected what would you expect to happen in time? Explain your answer.

6 How do you think that the bicycle in Figure 10.4 has been designed so that metal corrosion will not occur?

Factors that speed up rusting

Like most chemical reactions, rusting speeds up if the reactants are warmed. The presence of salt in the water on the metal also speeds up rusting.

Rust prevention

Rust can be prevented by keeping oxygen away from the iron or steel surface. This can be done by painting the surface or covering it in oil. However, if the paint becomes chipped or the oil is allowed to dry up, rust can begin to form. Steel can also be protected by covering the surface with chromium in a process called chromium plating. Steel used for canning foods is coated in a thin layer of tin.

The steel used for girders to build office blocks and bridges is coated in zinc in a process called galvanising or zinc plating.

Steel can also be prevented from rusting by mixing it with nickel and chromium to make the metal alloy called stainless steel. This is used for cutlery and kitchen sinks.

Figure 10.4 A bicycle with a frame which is resistant to rusting.

Other metals and the air

When aluminium or zinc is exposed to the air, the metal on its surface reacts with oxygen in the air and forms oxides. These oxides do not flake like rust but form a protective surface on the metal.

Zinc is used to protect iron because it is more reactive with oxygen than iron. Even if the zinc coating on galvanised iron or steel is chipped, the oxygen still reacts with the zinc instead of the exposed iron and rusting is prevented.

CORROSION

A black coating forms on the surface of silver exposed to the air. When this happens the silver is said to be tarnished. The tarnish can be removed by polishing.

The surface of bronze, which is an alloy of tin and copper, develops a thin green film over a long time. This coating is called patina.

Copper is used as a roofing material. Over time its surface reacts with air to form a green coating called verdigris. This substance contains copper sulphate.

Figure 10.5 A bronze statue with patina.

Figure 10.6 This copper roof is covered in verdigris.

♦ SUMMARY ♦

- Fast reactions can cause explosions (*see page 133*).
- Slow chemical reactions between oxygen and food make the food go bad (*see page 133*).
- Rusting is a slow reaction between iron and oxygen (*see page 134*).
- Rust weakens iron and steel (*see page 134*).
- Warmth and salt water speeds up rusting (*see page 135*).
- Rust can be prevented by keeping oxygen away from iron and steel (*see page 135*).
- Coatings form on some metals when they are exposed to the air (*see pages 135–136*).

End of chapter question

Anita and Arif have identical new bicycles. Anita lives in a dry, cool grassland area and Arif lives in a warm, wet region on the coast. Which bicycle may show signs of rusting first? Explain your answer.

11 Patterns of reactivity

Figure 11.1 Sodium under oil.

Sodium is so reactive that it has to be kept under a layer of oil. The reason for this is that it will readily react with water and oxygen in the air and produce flames. Sodium is a very soft metal and can be cut with a knife. If a small piece is cut and exposed to the air, its shiny surface quickly tarnishes as it reacts with the air.

In Chapter 10 iron was found to react with damp air and turn into rust. This reaction takes place much more slowly than the reaction between sodium and the air and suggests that metals have different speeds at which they react with other chemicals.

Reaction with oxygen

Here are some descriptions of the reactions that take place when certain metals are heated with oxygen.

Copper develops a covering of a black powder without glowing or bursting into flame. Iron glows and produces yellow sparks; a black powder is left behind. Sodium only needs a little heat to make it burst into yellow flames and burn quickly to leave a white powder behind (see Figure 11.2). Gold is not changed after it has been heated and then left to cool.

1 Arrange the metals mentioned in this section in order of their reactivity with oxygen. Start with what you consider to be the most reactive metal.

Figure 11.2 Sodium burning in a gas jar of oxygen (left). Sodium oxide powder is left behind (right).

PATTERNS OF REACTIVITY

Displacement reactions

When metals react with acids, they displace hydrogen from the acid and form a salt solution. In a similar way, a more reactive metal can displace a less reactive metal from a salt solution of the metal.

When a copper wire is suspended in a solution of silver sulphate, the copper dissolves into the solution to form copper sulphate and silver metal comes out of the solution and settles on the wire (see Figure 11.3).

Figure 11.3 Copper wire coils in silver sulphate solution. Silver is formed on the wire.

2 From the information about these two displacement reactions, arrange the three metals (copper, silver and iron) in order of reactivity – starting with the most reactive.

If an iron nail is placed in copper sulphate solution, the iron dissolves to form a pale green iron sulphate solution and the copper comes out of the solution and coats the nail (see Figure 11.4).

Figure 11.4 This iron nail has been left in copper sulphate solution. Copper has formed on the nail.

PATTERNS OF REACTIVITY

Look at Table 11.1 below to answer these questions.

3 How do you think the reactions that zinc makes with oxygen, water and acid compare with those that iron makes?

4 From Table 11.1, would you expect zinc to displace:
   a) iron,
   b) sodium,
   c) gold
   in displacement reactions? Explain your answer.

The reactivity series

The reactivity series is a list of metals arranged in order of their reactivity, starting with the most reactive. The series is produced by studying the reactions of metals with oxygen, water, hydrochloric acid and solutions of metal salts. Table 11.1 shows seven metals in the reactivity series and summarises their reactions with oxygen, water and hydrochloric acid.

Table 11.1 The reactivity series.

Metal	Reaction with oxygen	Reaction with water	Reaction with acid
sodium	oxide forms very vigorously	produces hydrogen with cold water	violent reaction
magnesium		produces hydrogen with steam	rate of reaction decreases down the table
zinc			
iron	oxide forms slowly		
copper	oxide forms without burning	no reaction with water or steam	very slow reaction
silver	no reaction		no reaction
gold			

5 Make a list of the metals in the reactivity series (Table 11.1). Then find the dates when the metals were discovered by examining all the information in the section on discovery of the elements on page 75. Explain the difference in the times of discovery by considering the reactivity of the metals.

Reactivity and metal extraction

The least reactive metals can be found in their pure state in the ground. More reactive elements combine with other elements and form compounds. The least reactive of these reactive metals can be released from their compounds by heat. The more reactive of these reactive metals combine more strongly with other elements and are separated from them by electricity.

PATTERNS OF REACTIVITY

♦ SUMMARY ♦

- Metals react at different speeds with oxygen (*see page 137*).
- Metals can be arranged in order according to how they react with oxygen, water and acid (*see page 138*).
- A more reactive metal can displace a less reactive metal from its salt (*see page 138*).
- Metals can be arranged in the order of their reactivity. This is called the reactivity series (*see page 139*).
- The reactivity of a metal affects the methods used to extract it from its ore (*see page 139*).

End of chapter question

Why do you think that you can find silver on its own in rocks but calcium is combined with other elements to make compounds such as calcium carbonate?

12 Preparing salts

The preparation of salts is described in Chapter 7. In this chapter the information is brought together for revision questions. However there are also questions for you to answer which test your knowledge of other aspects of chemistry.

The preparation of salts can be summarised by these three general word equations:

acid + metal → salt + hydrogen
acid + carbonate → salt + carbon dioxide + water
acid + base → salt + water

Preparation of zinc chloride

Granulated zinc is added to hydrochloric acid in a flask (see page 87). Bubbles of gas rise from the metal and pass through the liquid and escape into the air. Eventually the bubbles are no longer produced and some metal remains in the flask.

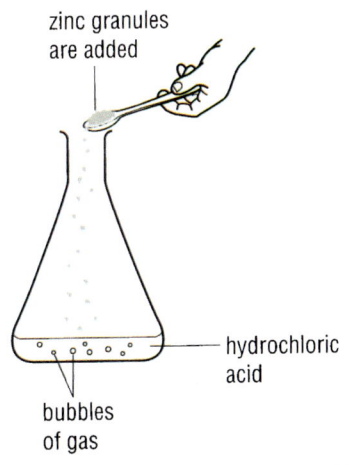

Figure 12.1 Adding zinc to hydrochloric acid.

The contents of the flask are poured into filter paper in a filter funnel. The zinc metal remains behind and the liquid passes through and falls into a beaker. The liquid is poured into an evaporating dish and heated until some solid appears. The mixture is then left to cool and more evaporation takes place. After this the mixture is filtered again.

---

1 Name an organic acid and a mineral acid.
2 What colours would you expect acids to turn universal indicator paper?
3 An alkali is a soluble base. What colour would you expect alkalis to turn universal indicator paper?
4 How could you test a gas to see if it was:
   a) hydrogen,
   b) carbon dioxide?
5 Why do you think that granulated zinc is used instead of a block? You may have to look at page 154 to help you answer.
6 What passed through the filter paper when the flask was emptied?
7 Why was the solution heated before it was left in an evaporating dish?
8 Write the word equation for this reaction.
9 Zinc sulphate can be prepared in a similar way. Write a word equation for the reaction.

141

PREPARING SALTS

Preparation of calcium chloride

Some marble chips are added to hydrochloric acid in a flask. Bubbles are produced and the chips dissolve. Some more chips are added and more bubbles are produced and then the reaction stops. Some of the chips are left in the solution.

The contents of the flask are poured into a filter paper in a filter funnel and the solution and chips are separated.

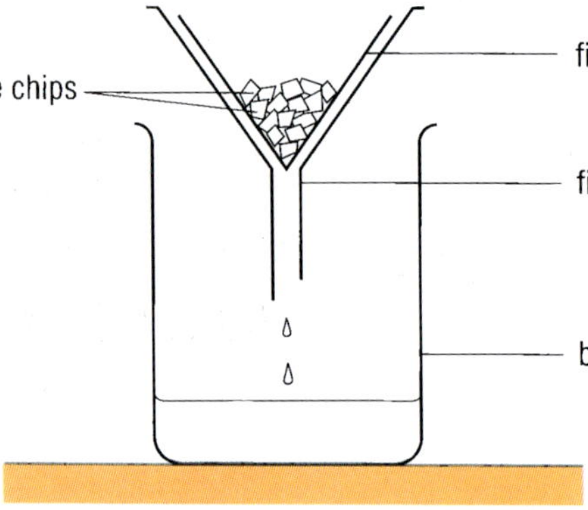

Figure 12.2 Filtering out marble chips using a filter funnel.

The liquid is poured into an evaporating dish and heated until some solid appears.

The mixture is then left to cool and more evaporation takes place. When the mixture has been left to cool it is filtered again.

10 Why did all the chips added at first dissolve?
11 Why did some of the chips added later not dissolve?
12 Why were the contents of the evaporating dish filtered?
13 What compound is marble made from?
14 Write the word equation for the reaction.

Preparation of copper sulphate

Some copper oxide is added to dilute sulphuric acid. The mixture is warmed and stirred. When all the copper oxide has dissolved, some more is added and the warming and stirring is continued until some copper oxide remains undissolved.

The undissolved copper oxide is removed by filtration and the filtered liquid is poured into an evaporating dish and heated until some solid appears.

The mixture is then left to cool and more evaporation takes place. When the mixture has been left to cool it is filtered again.

15 Why was heat used in the preparation of copper sulphate?
16 What was the effect of stirring the mixture?
17 Write the word equation for the reaction.

Preparation of potassium chloride

The preparation of sodium chloride is given in detail on page 85. Here is a summary of the preparation of potassium chloride which is prepared in a similar way.

PREPARING SALTS

18 How is the preparation given here different from the preparation of sodium chloride?
19 Why is an indicator used?
20 Why is the flask swirled?
21 What colour is litmus in:
   a) an alkaline solution,
   b) an acid solution?
22 Why was the colour removed?
23 How could the preparation be carried out without the use of charcoal?
24 Write the word equation for the reaction.
25 How is lead iodide different from silver chloride?
26 Why does the lead iodide not pass through the filter paper?
27 Write a word equation for the reaction.

However there are some differences that you might like to compare (see question 18).

A volume of potassium hydroxide is measured out in a measuring cylinder and poured into a flask. A few drops of litmus solution are added to the flask and the flask is swirled. The flask is placed under a burette containing hydrochloric acid and the acid is added gradually, with a swirling of the flask taking place after each addition. When the litmus indicator changes colour no more acid is added. Charcoal is added to absorb the colour of the litmus and the mixture is filtered.

The filtered liquid is poured into an evaporating dish and heated until some solid appears.

The mixture is then left to cool and more evaporation takes place. When the mixture has been left to cool it is filtered again.

Preparation of lead iodide

If potassium iodide is added to lead nitrate a yellow precipitate of lead iodide is produced. The solution can be filtered. This leaves the lead iodide in the filter paper. The lead iodide can then be left to dry.

♦ SUMMARY ♦

- The preparation of salts can be summarised by three word equations (*see page 141*).
- Zinc chloride is prepared from granulated zinc and hydrochloric acid (*see page 141*).
- Calcium chloride is prepared from marble chips and hydrochloric acid (*see page 142*).
- Copper sulphate is prepared from copper oxide and sulphuric acid (*see page 142*).
- Potassium chloride is prepared from potassium hydroxide and hydrochloric acid (*see pages 142–143*).
- Lead iodide is prepared from potassium iodide and lead nitrate (*see this page*).

End of chapter question

What are the two physical processes used in the separation of substances in preparing salts? How does each process bring about separation?

13 Exothermic and endothermic reactions

An exothermic reaction is one in which heat is given out. An endothermic reaction is one in which heat is taken in.

Exothermic reactions

The most familiar exothermic reaction to everyone is combustion. When a flame develops in this reaction, combustion is then called burning. We say that a burning substance is on fire.

Figure 13.1 Food is being cooked here on an open fire. The heat from this exothermic reaction is used to bring about non-reversible changes in the food. We recognise these changes by saying that the food is cooking.

Burning

Many substances are burned to provide heat or light. They are called fuels. Wood, coal, coke, charcoal, oil, diesel oil, petrol, natural gas and wax are examples of fuels. The heat may be used to warm buildings, cook meals, make chemicals in industry, expand gases in vehicle engines and turn water into steam to drive generators in power stations. Some gases and waxes are used to provide light in caravans and tents.

Natural gas is an example of a hydrocarbon. It is made of carbon and hydrogen. When natural gas burns, carbon dioxide and water (hydrogen oxide) are produced. Many other fuels such as coal, coke and petrol contain hydrocarbons (see page 105).

Investigating a burning candle

A candle can be used to investigate how fuels burn.

1 Give a use for each of the fuels listed in the paragraph on burning. How many different uses can you find?

EXOTHERMIC AND ENDOTHERMIC REACTIONS

Investigation 1

If a burning candle is put under a thistle funnel which is attached to the apparatus shown in Figure 13.2 and the suction pump is switched on, a liquid collects in the U-tube and the lime water turns cloudy.

When the liquid is tested with cobalt chloride paper, the paper turns from blue to pink. This shows that the liquid is water. The cloudiness in the lime water indicates that carbon dioxide has passed into it.

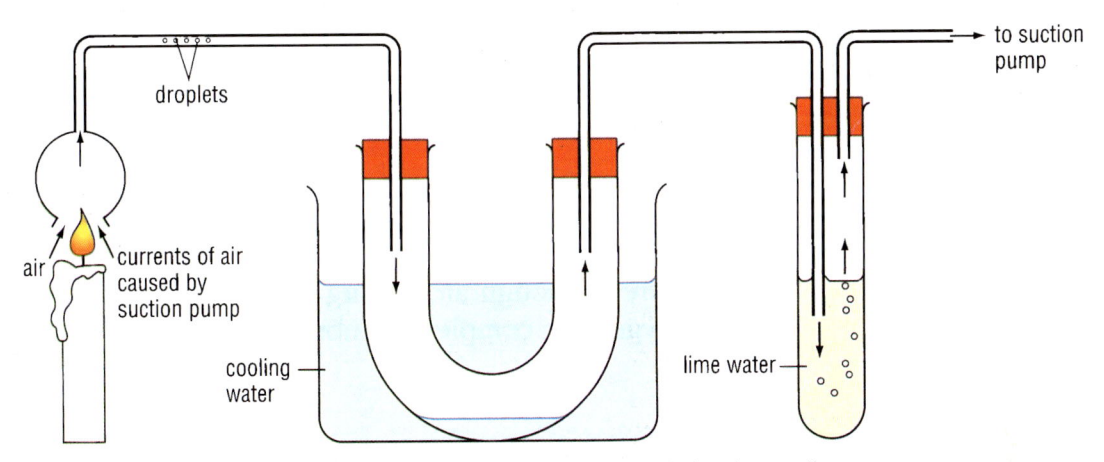

Figure 13.2 Testing the products of a burning candle.

Investigation 2

If a beaker is placed over a burning candle, the candle will burn for a while and then go out. A change has taken place in the air that makes it incapable of letting things burn in it.

The test for oxygen is made by plunging a glowing splint of wood into the gas being tested. If the gas is oxygen, the splint bursts into flame. When air from around the burned-out candle is tested for oxygen, the glowing splint goes out. This indicates that oxygen is no longer present. The oxygen in the air under the beaker has been used up by the burning candle.

From the information provided by these two investigations with candles, the following word equation can be set out:

hydrocarbon + oxygen → carbon dioxide + water

Natural gas is a hydrocarbon called methane. When it burns, it breaks down exactly like the hydrocarbons in candle wax. The word equation for this reaction is:

methane + oxygen → carbon dioxide + water

Both of these word equations are examples of complete combustion. This only happens when there is enough oxygen available.

EXOTHERMIC AND ENDOTHERMIC REACTIONS

The danger of incomplete combustion

If there is insufficient oxygen to support complete combustion, incomplete combustion takes place. Carbon monoxide is a very dangerous chemical produced by incomplete combustion. It is produced instead of carbon dioxide. Carbon monoxide is produced in car engines and is released in the exhaust fumes.

Incomplete combustion also occurs when a gas fire has been incorrectly fitted and cannot draw enough oxygen from the room it is heating. Carbon monoxide is a colourless, odourless gas so you do not know when it is being produced. If it is breathed in, it stops the blood taking up oxygen and circulating it round the body. People have died from breathing carbon monoxide from badly fitted fires. All gas fires must be fitted by a trained engineer and used in a well ventilated room so that there is enough air passing through the fire to provide oxygen for complete combustion of the gas.

2 What happens to the carbon in natural gas when the gas burns in a badly fitted gas fire? Explain your answer.

3 Which fuels would you take on a camping trip? Explain your choices.

Improving efficiency

Wood was the first fuel. In many parts of the world it is still used as a fuel today.

It is often in short supply so ways have been developed to use the fuel more efficiently. Figure A shows a stove that has been developed in Sri Lanka to provide heat for cooking two pots for a meal at once. One pot is used to boil rice while the other is used to cook vegetables.

The damper can be raised or lowered and used to control the amount of air reaching the fire. This in turn controls the burning of the wood. The baffle is used to control the direction of the flames. They can be made to go straight up and heat the pot above them.

1 How will designing more efficient stoves help conserve fuel?

2 How will using the damper make the wood burn more slowly or more quickly?

3 In the past people used to put a pan on three stones over a fire. Why is the stove in Figure A an improvement?

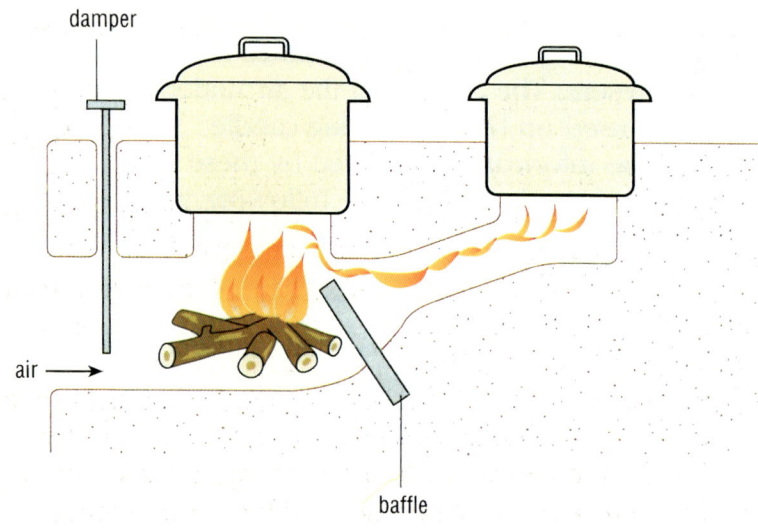

Figure A Heating two pots at once.

EXOTHERMIC AND ENDOTHERMIC REACTIONS

Measuring energy in a fuel

The energy in a fuel is found by collecting the heat energy released from a burning fuel and measuring it. The device which is used to do this is called a calorimeter.

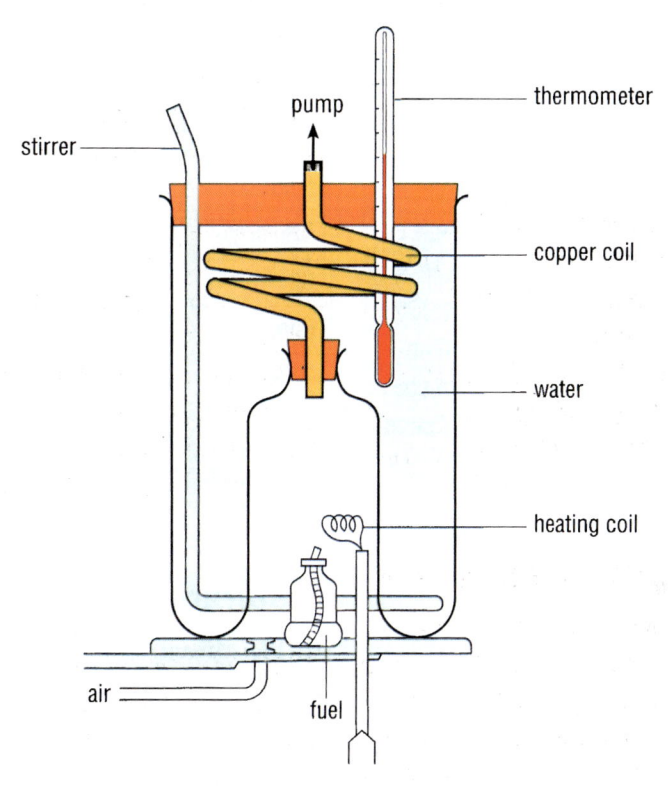

Figure 13.3 A calorimeter.

The calorimeter consists of a chamber in which the fuel is burnt, a water jacket around the chamber to collect the heat, and a system of tubes which bring air to the fuel and carry away the hot gases so that they release their heat into the water jacket.

The mass of the fuel is found before it is placed in the calorimeter. The temperature of the water is taken before the fuel is ignited. Air passes over the fuel as it burns and the water in the jacket is stirred. The temperature of the water is taken until the fuel burns out.

It has been found that 4.2 kilojoules of energy raises the temperature of 1000 g of water by 1 °C. If the volume of water in the calorimeter is 500 g, 2.1 kJ will raise the temperature by 1 °C. The amount of energy released by the fuel is found by multiplying the rise in temperature by 2.1.

An alternative way to find the energy released by a fuel is to repeat the experiment with a wire heating element in the chamber. Electricity is passed through a

4 Why is air pumped through the chamber instead of just letting the fuel use the air that is present?
5 Why is the pipe carrying the hot gases made of a copper coil?
6 Why is the water stirred?
7 How do you find the rise in temperature of the water?
8 If a calorimeter had 500 g of water in its jacket and the water temperature rose by 25 °C, how many joules of energy were released by the fuel?
9 If the fuel had a mass of 2 g, how much heat energy does each gram of fuel release?

147

EXOTHERMIC AND ENDOTHERMIC REACTIONS

joulemeter and the element, and the temperature is allowed to rise until it reaches the same temperature as that when the fuel burnt out. The electricity is then switched off and the joulemeter is read to find the amount of energy released.

The energy in different fuels has been measured and the results are shown in Table 13.1.

Table 13.1 Energy in various fuels.

Fuel	Energy kJ/g
Natural gas	56
Oil	48
Coal	34
Coke	32
Wood	22
Carbohydrates (e.g. starch and sugars)	16

10 How many grams of biscuit (carbohydrates) would you need to provide you with the same amount of energy as 100 g of coke?

11 Bjorn and Ingrid live in a country with cold winters. They live in identical houses but Bjorn heats his house with coal while Ingrid heats her house with wood. They both fill up their fuel stores for the winter. Who will need the larger store? Explain your answer.

Respiration

All life processes require energy. Living things obtain their energy from their food. The process in which energy is released from food is called respiration. In this process a food substance called glucose reacts with oxygen. The word equation for this reaction is:

glucose + oxygen → carbon dioxide + water

The energy released in respiration is used to make substances inside the body and to make a body move. Some of the energy is released as heat (see Figure 13.4).

12 Is food a fuel? Explain your answer.

13 Is respiration like burning? Explain your answer.

Figure 13.4 This runner has been generating a great deal of heat.

EXOTHERMIC AND ENDOTHERMIC REACTIONS

Endothermic reactions

Photosynthesis

Figure 13.5 Photosynthesis takes place in plant leaves and is a reaction that takes in energy from sunlight.

The food used by living things is manufactured by plants. Plants do not release energy as they make food, they take it in – from sunlight. This food-making process is called photosynthesis. Carbon dioxide and water are used to trap the energy and make glucose. The word equation for this reaction is:

carbon dioxide + water → glucose + oxygen

Sherbet

Sherbet is a popular sweet. It is made from citric acid and sodium hydrogencarbonate. When you put sherbet in your mouth, it feels cool. This is due to an endothermic reaction taking place which results in heat being taken from your body. The reaction occurs when the sherbet dissolves in your saliva and the two chemicals react together. The word equation for the reaction is:

citric acid + sodium hydrogencarbonate → sodium citrate + carbon dioxide + water

Figure 13.6 Sherbet makes your mouth feel cool.

14 What makes the sherbet fizz?

149

EXOTHERMIC AND ENDOTHERMIC REACTIONS

Decomposition of limestone

Limestone is heated to produce lime (see page 95). When the limestone is heated it decomposes to form calcium oxide and carbon dioxide. Heat is needed for this reaction. When heat is removed from the limestone, it stops decomposing.

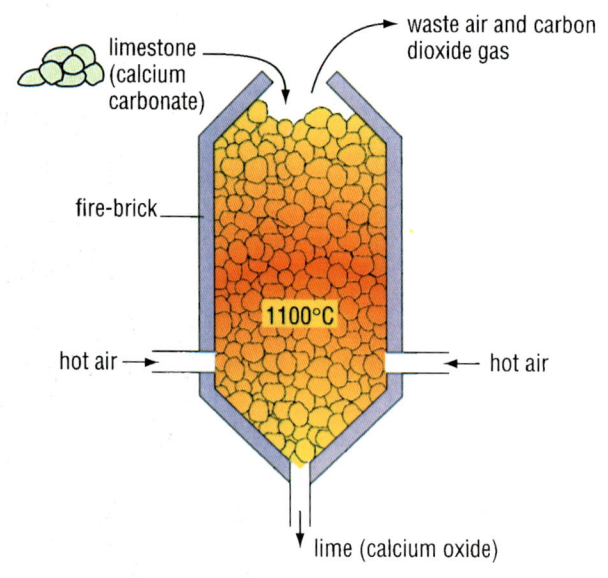

Figure 13.7 A lime kiln.

Large amounts of limestone are converted into lime in a lime kiln (see Figure 13.7). Small limestone rocks are poured into the top of a kiln, which is then sealed. Heat from gas burners decomposes the limestone. Streams of air entering the bottom of the kiln carry away carbon dioxide from the top of the kiln and prevent it reacting with the calcium oxide. If the carbon dioxide did react with the calcium oxide, calcium carbonate would form again.

15 If air was not allowed to stream through the kiln, how would the production of lime be affected? Explain your answer.

Cooking

The reversible changes which take place during cooking are due to endothermic reactions. Energy from a heat source, such as an oven, is taken up by the food and used in chemical reactions to cook it (see page 144).

Digestion

We need energy from our food to keep us alive. The energy is released in respiration and used in life processes. One of the life processes is digestion in which food is broken down by chemical reactions involving enzymes.

EXOTHERMIC AND ENDOTHERMIC REACTIONS

Mammals and birds conserve the heat released by respiration and keep a constant body temperature. This means that heat is always available for the endothermic reactions of digestion. Reptiles, such as crocodiles and lizards, do not keep a constant body temperature. It varies with their surroundings. In order to get the heat they need for their body processes reptiles may rest in the sunshine. As they warm up their digestive processes can speed up and digest the contents of their stomachs.

Figure 13.8 Sunbathing crocodiles.

♦ SUMMARY ♦

- Burning is an exothermic reaction (*see page 144*).
- Incomplete combustion is dangerous (*see page 146*).
- The energy in a fuel can be measured (*see page 147*).
- Respiration is an exothermic reaction (*see page 148*).
- Photosynthesis is an endothermic reaction (*see page 149*).
- The chemicals in sherbet take part in an endothermic reaction (*see page 149*).
- Limestone is decomposed in an endothermic reaction (*see page 150*).
- Digestion is an endothermic reaction (*see page 150*).

End of chapter questions

1. Plan an investigation to compare the amount of energy in two fuels.
2. What kinds of reactions, exothermic or endothermic, take place when:
    - **a)** a plant makes food in sunlight,
    - **b)** wood is used to cook food,
    - **c)** the food in the plant is cooked on the fire,
    - **d)** the food is digested in your body,
    - **e)** some of the digested food is used in respiration?

14 Rates of reaction

Chemical reactions take place at different speeds. The chemical reaction which blows rock out of a cliff face in a quarry is a very fast reaction.

Figure 14.1 An explosion in a quarry.

Some reactions, such as the setting of concrete, are very slow and may take two days or more to complete.

Figure 14.2 The setting of concrete takes two days.

In chemistry the speed of a reaction is studied by considering the rate at which the chemicals in the reaction change. Rate is a measure of the change in a certain amount of time. The rate may show how fast the mass of the reactants changes in a certain amount of time, or how fast a product is produced in a certain amount of time.

RATES OF REACTION

Change in mass of reactants

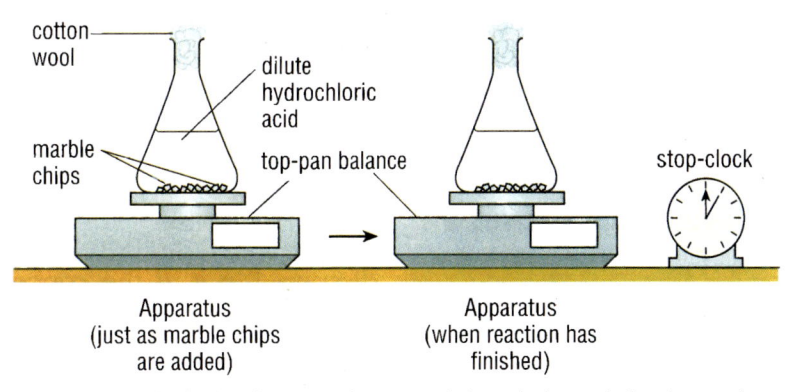

Figure 14.3 Weighing the mass of reactants before, during and after the reaction.

After the mass of the reactants (marble chips and hydrochloric acid) has been recorded, the reactants are mixed together in the flask and their mass recorded (see Figure 14.3). The mass is then recorded regularly over a certain number of minutes. The word equation for the reaction is:

calcium + hydrochloric → calcium + carbon + water
carbonate   acid   chloride   dioxide

The carbon dioxide escapes from the flask and accounts for the change in mass.

Change in volume of a product

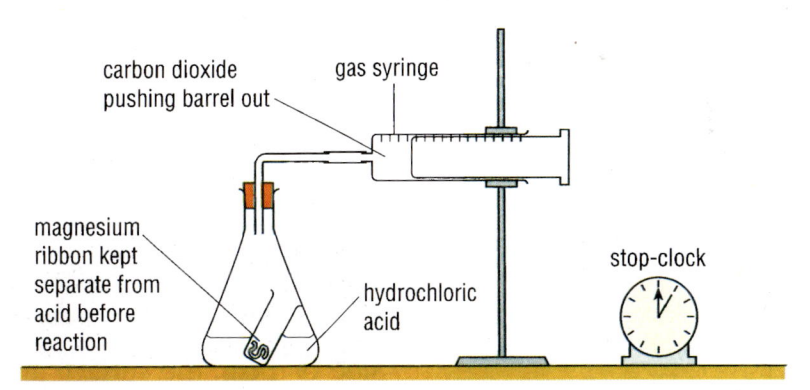

Figure 14.4 The volume of hydrogen produced can be measured.

When the reactants (magnesium and hydrochloric acid) are mixed together hydrogen is produced as the word equation describes:

magnesium + hydrochloric → magnesium + hydrogen
                acid   chloride

153

RATES OF REACTION

1 In a model volcano some vinegar was poured onto sodium bicarbonate. A reaction took place which produced a fizzy liquid which flowed down the sides of the volcano like lava. If water is added to the vinegar and the reaction is repeated, will the eruption of the volcano be stronger or weaker? Explain your answer.

As the gas is produced it pushes the plunger in the syringe to the right and the volume produced every minute can be measured.

The effect of concentration

The concentration of a solution is a measure of the solute dissolved in it. A solution of low concentration has a small amount of solute dissolved in it. A solution of high concentration has a large amount of solute dissolved in it. If the concentration of a reactant is increased the rate of reaction increases. If the concentration of one reactant is doubled the rate of the reaction may be doubled.

The effect of particle size

On page 133 the explosion in a mine was accounted for by the small size of the particles of coal dust. This suggests that particle size affects the rate of reaction. When a solid, such as a piece of coal, breaks up into smaller particles its surface area increases as the following example shows. Imagine a cube-shaped piece of coal with sides 2 cm long. It has six surfaces. Each one is $2 \times 2 = 4 \, cm^2$ in area. The surface area of the cube is $6 \times 4 = 24 \, cm^2$.

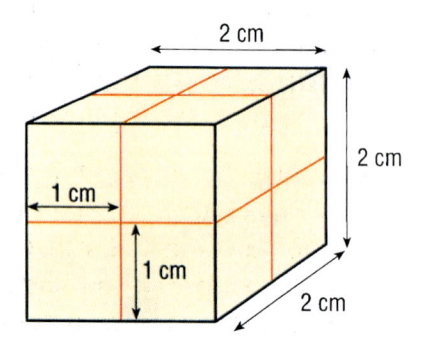

Figure 14.5 One large cube can be cut into eight smaller cubes.

If the cube was broken into eight cubes, each with a side of 1 cm, the surface area of each cube would be $6 \times 1 \times 1 = 6 \, cm^2$. As there are now eight cubes their total surface area is $8 \times 6 = 48 \, cm^2$. The surface area has doubled. If the eight cubes were broken up into smaller pieces the surface area would also increase.

The surface area of a reactant is its point of contact with other reactants. If the surface area is increased, the reactants can come into greater contact and the reaction rate will increase.

RATES OF REACTION

2 The graph in Figure 14.6 shows the volume of carbon dioxide produced when large marble chips take part in a reaction with hydrochloric acid.
   a) Make a copy of the graph and draw in the line you would expect to see when smaller chips are used.
   b) Explain your answer.

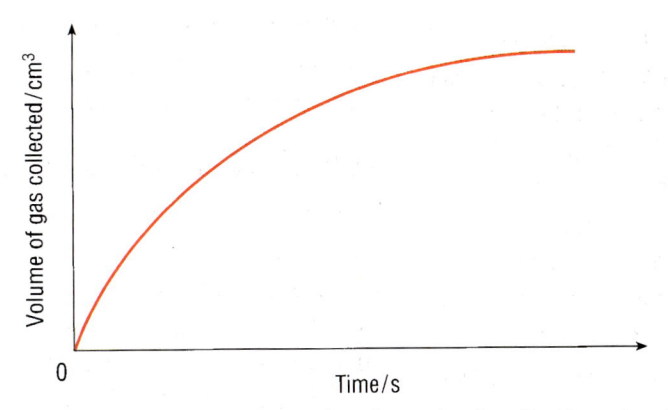

Figure 14.6 A graph showing the volume of carbon dioxide produced over time.

3 Two cartons of milk were opened and one was left by a radiator while the other was placed in a fridge. How will the milk in the two cartons differ after 3 days? Explain your answer.

The effect of temperature

The rate of reaction increases if the temperature is raised. The rate decreases if the temperature is lowered. If the temperature of the reactants is raised by 10°C the speed of the reaction may be doubled.

Measuring the effect of temperature

Sodium thiosulphate is a substance which dissolves in water. When hydrochloric acid is added to a solution of sodium thiosulphate, sodium chloride, water, sulphur and sulphur dioxide are produced.

The sulphur forms a precipitate which clouds the solution and the speed at which this cloudiness appears can be used as a measure of the rate of the reaction. An investigation takes place in the following way.

1 A flask containing the reactants is placed over a piece of paper with a cross on it.
2 The reactants are viewed from the top of the flask and a stop-clock is started.
3 When the sulphur precipitate clouds the solution so much that the cross cannot be seen, the stop-clock is stopped and the time is recorded (see Figure 14.7).

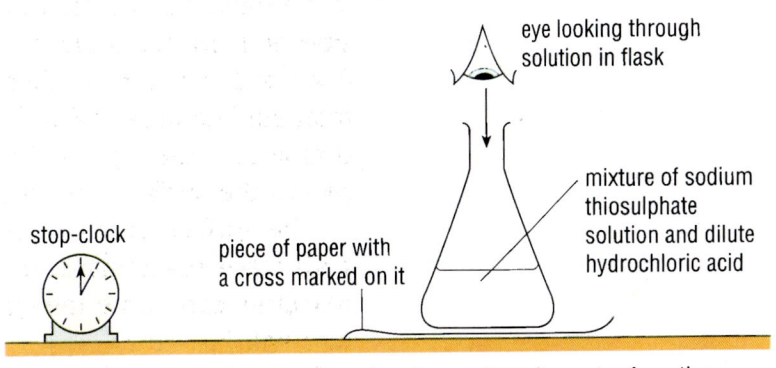

Figure 14.7 An experiment to assess how temperature alters rate of reaction.

RATES OF REACTION

4 a) Plot the data in the table onto a graph.
  b) What is the shape of the graph?
  c) What would you predict the speed of the reaction to be at 32.5 °C?
  d) What would you predict the temperature to be if the reaction took 76 seconds?

5 Write a word equation for the reaction between sodium thiosulphate and hydrochloric acid.

If the experiment is repeated several times with the reactants at increasingly higher temperatures, a table of data can be produced as Table 14.1 shows.

Table 14.1 Reaction time at different temperatures.

Temp °C	Time of reaction in secs
25	110
30	80
35	60
40	46
45	38
50	30

A fuel for the future?

When hydrogen gas is introduced, as a jet, into air or oxygen, it burns steadily. The reaction produces a great deal of heat energy. In rockets the heat energy is used to produce a large amount of fast moving gases which rush out at the end of the rocket engine. The force of the moving gas is balanced by a force in the opposite direction, which lifts the space craft into the sky.

Catalysts

A catalyst is a substance which is added to the reactants to increase the rate at which they react. At the end of the reaction the catalyst remains unchanged chemically. This means that it can be used again. Only a small amount of catalyst is needed to produce a large increase in the rate of a reaction. A catalyst usually only works at speeding up one reaction. It does not speed up a range of reactions.

Using a catalyst in the laboratory

Hydrogen peroxide is a liquid which very slowly breaks down into water and oxygen. However the rate of reaction can be greatly increased by adding a small amount of manganese dioxide. The reaction becomes so fast that the liquid fizzes as the oxygen escapes.

RATES OF REACTION

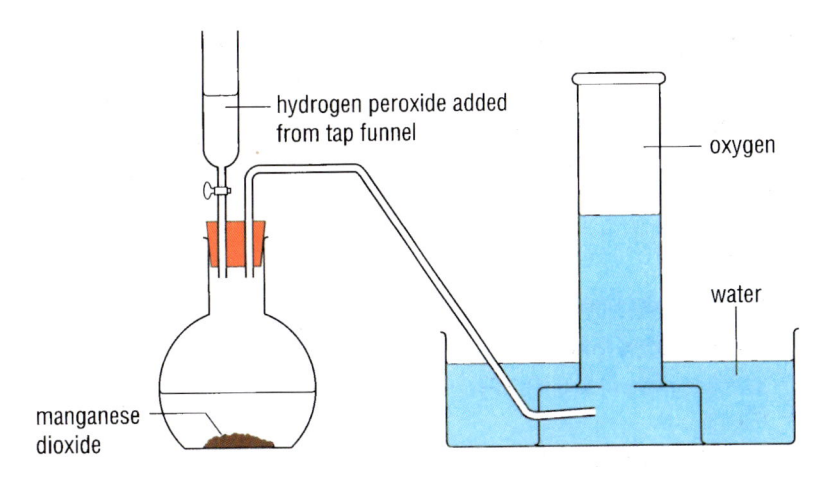

Figure 14.8 Hydrogen peroxide breaks down more quickly when manganese dioxide is added to it.

Using catalysts in industry

Catalysts are used in the making of sulphuric and nitric acids.

Sulphuric acid

A catalyst made from vanadium oxide is used in the manufacture of sulphuric acid. It is used to speed up the reaction between sulphur dioxide and oxygen. Before the gases pass over the catalyst their pressure is increased to twice that of the atmosphere and they are heated to 400–500 °C. When they pass over the catalyst the gases react quickly to produce sulphur trioxide. This is dissolved in oleum (see page 89) to make sulphuric acid.

Nitric acid

Nitric acid is made by mixing ammonia with air. The process takes place in several stages during which the mixture of gases is heated to 900 °C and passed over a catalyst made from platinum and rhodium.

Nitric acid is used to make fertiliser, explosives such as trinitrotoluene (TNT), pharmaceuticals and synthetic fibres.

Catalysts and air pollution

Car engines produce a range of gases. Some of them such as carbon monoxide, nitrous oxides and hydrocarbons are harmful. The catalytic converter used on cars is a device which forms part of the exhaust system.

Inside the converter is a catalyst made of platinum and rhodium. The waste gases from the engine take part in chemical reactions in the converter which produce water, nitrogen and carbon dioxide.

6 What raw materials are used to make nitric acid?

7 How does nitric acid help in the blasting of rock out of the ground in a quarry?

157

RATES OF REACTION

8 The middle of a catalytic converter has a honeycomb structure. Why is this structure used?

9 In Figure 14.10 why does the rate of reaction increase with temperature on the left hand side of the graph?

10 In Figure 14.10 why does the rate of reaction decrease on the right hand side of the graph?

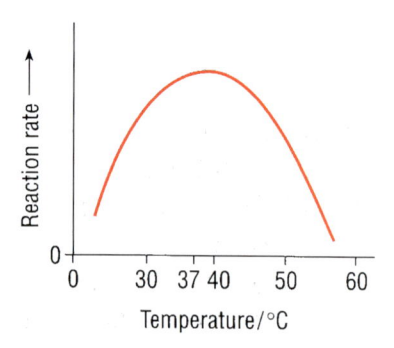

Figure 14.10 Graph showing how reaction rate varies with temperature.

11 In Figure 14.11 at what pH is the rate of reaction greatest?

12 Make a drawing of the graph in Figure 14.11 and add to it a line representing:
   a) the rate of a reaction due to an enzyme that works best in acidic conditions,
   b) the rate of a reaction due to an enzyme that works best in alkaline conditions.

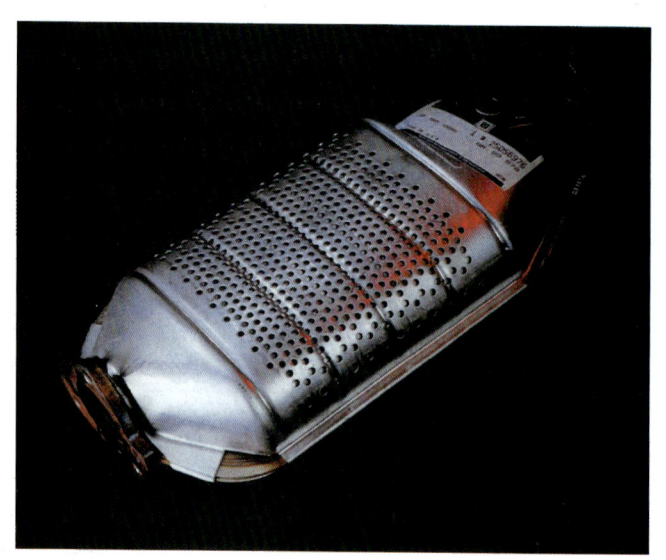

Figure 14.9 A catalytic converter.

Biological catalysts

Biological catalysts are called enzymes. They speed up the rate of reactions of life processes in plants and animals. Many chemical reactions take place in the liver and enzymes are present to speed them up. If a piece of liver is placed in hydrogen peroxide it speeds up its decomposition in a similar way to manganese dioxide. Enzymes are made from proteins. These substances are destroyed by heat. The graph in Figure 14.10 shows how the rate of a reaction catalysed by an enzyme varies with temperature.

The rate of a reaction catalysed by an enzyme is also affected by the pH of the liquid it is in.

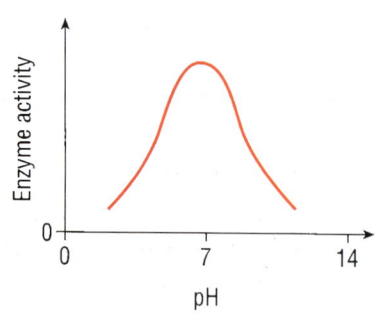

Figure 14.11 Graph showing how enzyme activity varies with pH.

In the human digestive system there are enzymes to break up proteins, fats and carbohydrates in food. Proteins and fats also form most of the stains on dirty clothing and biological washing powders containing enzymes have been developed to break them down. Biological washing powders are not suitable for

158

everyone as some people are allergic to the enzymes and develop a rash when they come into contact with them on clothing.

The kinetic theory and rates of reaction

In Chapter 4 matter was described as being made up from tiny particles, much smaller than the particles of coal described earlier in this chapter. The particle model of matter can also be used to explain the factors which affect the rates of reactions. Particles take part in reactions when they collide together so any factors which increase the chance of collisions will increase the rate of reaction.

Concentration

A concentrated solution has more particles in it that are available to react than a dilute solution. This means that increasing the concentration of a solution increases the number of particles and increases the number of collisions and the rate of reaction.

Particle size

Here the word 'particle' does not mean tiny particles but particles like dust and flour. On page 154 it was shown how a large cube had a smaller surface area than many smaller ones. Reactions take place at surfaces – so the smaller the surface area the less chance of collisions between the particles in the surface and the particles of the reactant in the liquid or gas next to the surface. Increasing the surface area increases the chance of more collisions and increases the rate of reaction.

Temperature

The speed at which particles move depends on their temperature. If their speed is increased, the chance of collision increases and rate of reaction increases.

Catalysts

Catalysts have a surface on which reactants can settle and join together. The catalyst increases the chance of the particles meeting and so increases the rate of reaction. Once the particles have met and joined together, they move away from the surface of the catalyst. This leaves room for other particles to join together.

RATES OF REACTION

♦ SUMMARY ♦

- The rate of reaction is a measure of the rate at which the chemicals in a reaction change (*see page 152*).
- Rate of reactions can be found by measuring the changes in the mass of the reactants (*see page 153*).
- Rate of reactions can be found by measuring changes in the volume of a product (*see page 153*).
- The rate of reaction is affected by the concentration of one reactant (*see page 154*).
- The rate of reaction is affected by the particle size of one reactant (*see page 154*).
- The rate of reaction is affected by the temperature of the reactants (*see page 155*).
- A catalyst is a substance that speeds up the rate of a reaction (*see page 156*).
- Biological catalysts are called enzymes (*see page 158*).
- The particle theory of matter can be used to explain why rates of reaction change (*see page 159*).

End of chapter question

Shazia has built a campfire and it is burning well. Robert collects some damp logs and puts them on the fire. Shazia is annoyed with Robert because the fire now burns more slowly. Why do you think that there has been a change in the rate of reaction?

15 The periodic table

In Chapter 6 the atoms of the 20 lightest elements were introduced. In this chapter we shall look at them again and how the elements are sorted out and arranged in the periodic table (see page 82). The chances are that there is a copy of this table on your chemistry laboratory wall.

> **For discussion**
>
> How many of the atoms in Table 15.1 can you recognise by their symbols? You can check your answer by looking at Table 6.2 on page 82.

The structure of atoms

Atoms are made from three sub-atomic particles – protons, neutrons and electrons (see page 80).

Atoms and electron shells

Table 15.1 Electron shells of the twenty lightest elements.

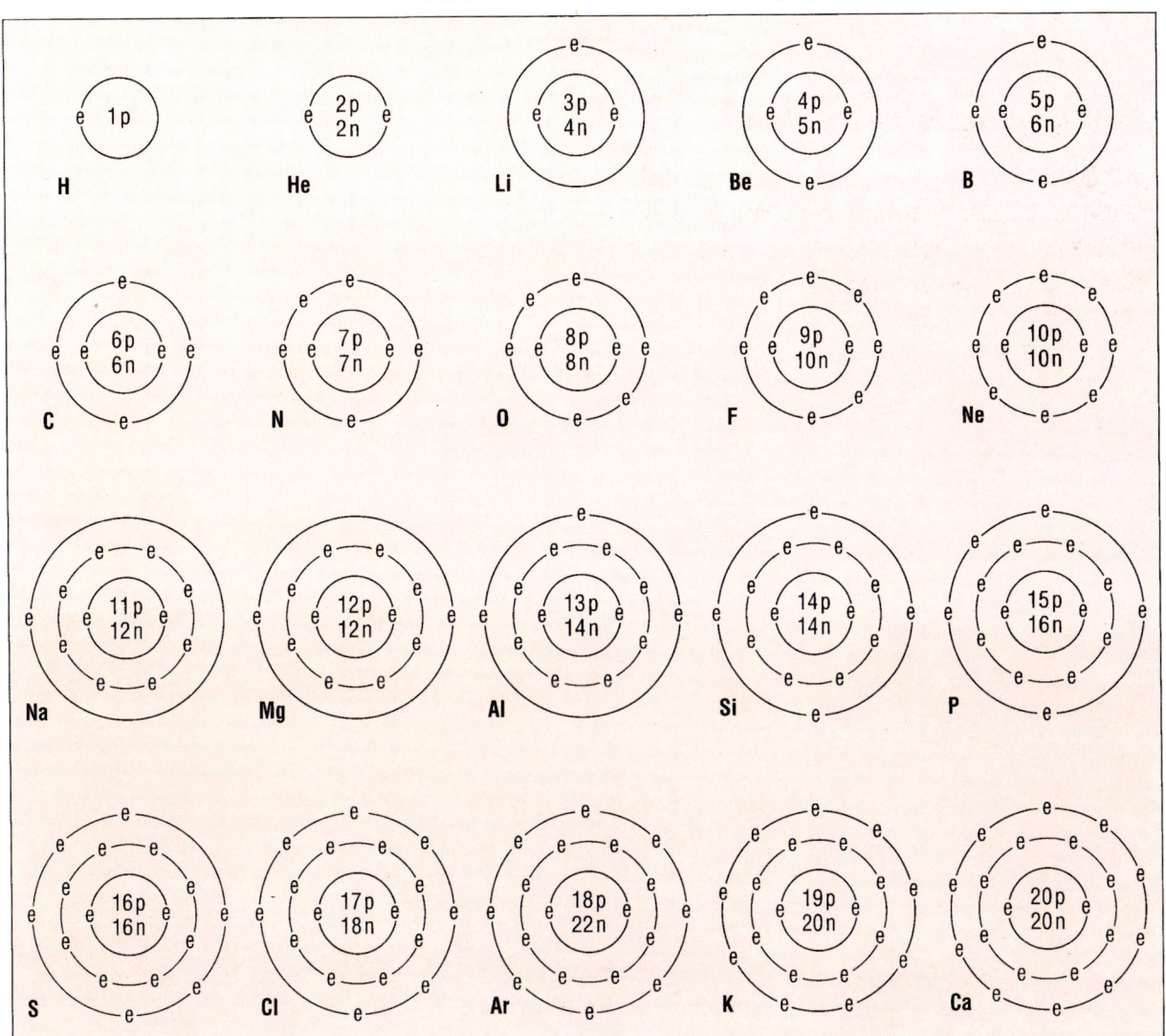

Element	Protons	Neutrons
H	1p	—
He	2p	2n
Li	3p	4n
Be	4p	5n
B	5p	6n
C	6p	6n
N	7p	7n
O	8p	8n
F	9p	10n
Ne	10p	10n
Na	11p	12n
Mg	12p	12n
Al	13p	14n
Si	14p	14n
P	15p	16n
S	16p	16n
Cl	17p	18n
Ar	18p	22n
K	19p	20n
Ca	20p	20n

161

THE PERIODIC TABLE

1 How many electrons are there in each shell of an atom of lead?

The electrons are arranged in groups at different distances from the nucleus. They are described as being arranged in shells. For example, the carbon atom has two electrons close to the nucleus making an inner shell and four electrons further away making an outer shell. Many atoms have more shells than this. For example, the lead atom has six shells (see Figure 15.1).

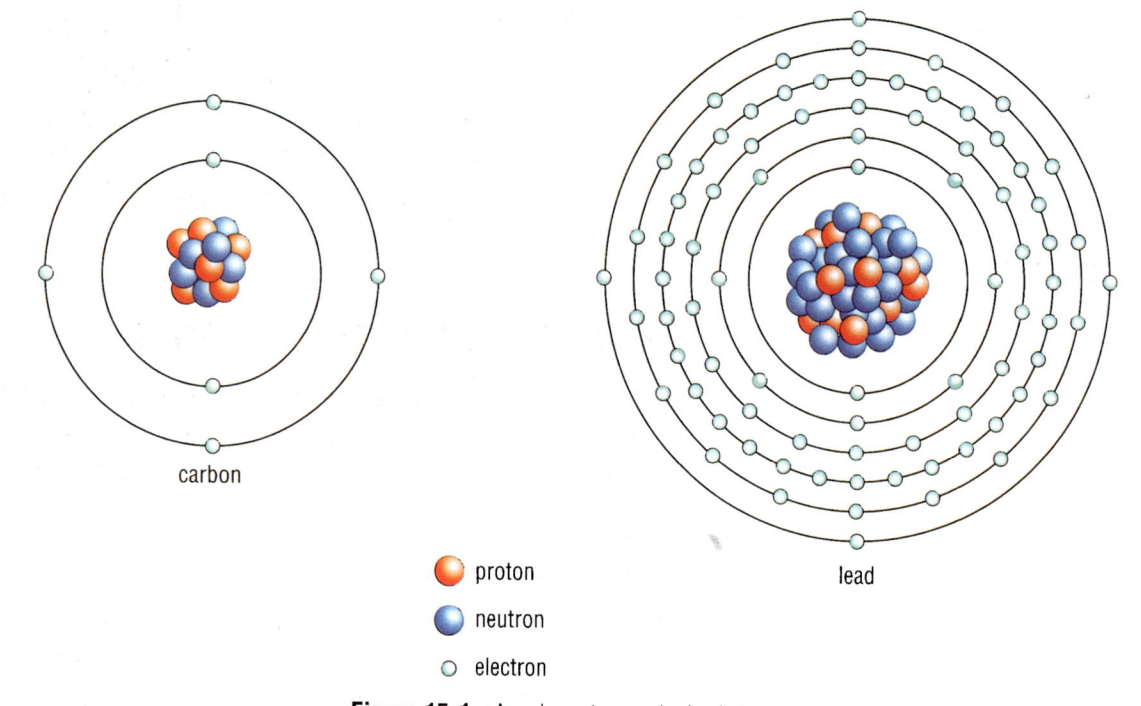

- proton
- neutron
- electron

Figure 15.1 A carbon atom and a lead atom.

2 What are isotopes?

All the atoms in each element have the same number of protons. For example, carbon atoms always have six protons and sodium atoms always have eleven protons.

The number of neutrons in the atoms of an element may vary. Most carbon atoms, for example, have six neutrons but about 1% of carbon atoms have seven neutrons and an even smaller amount of carbon atoms have eight neutrons. These atoms of an element that have different numbers of neutrons are called isotopes.

For discussion
Look back at how particles are arranged in solids, liquids and gases (page 55). How do you think atoms or groups of atoms (compounds) are arranged in the three states of matter?

Atomic number

In the nucleus of each atom of each element there is a certain number of protons. This number is different from the number of protons in the nuclei of any other element's atoms. The number of protons in an atom is called the atomic number. Elements are arranged in order of their atomic number in the periodic table (see Figure 15.2).

THE PERIODIC TABLE

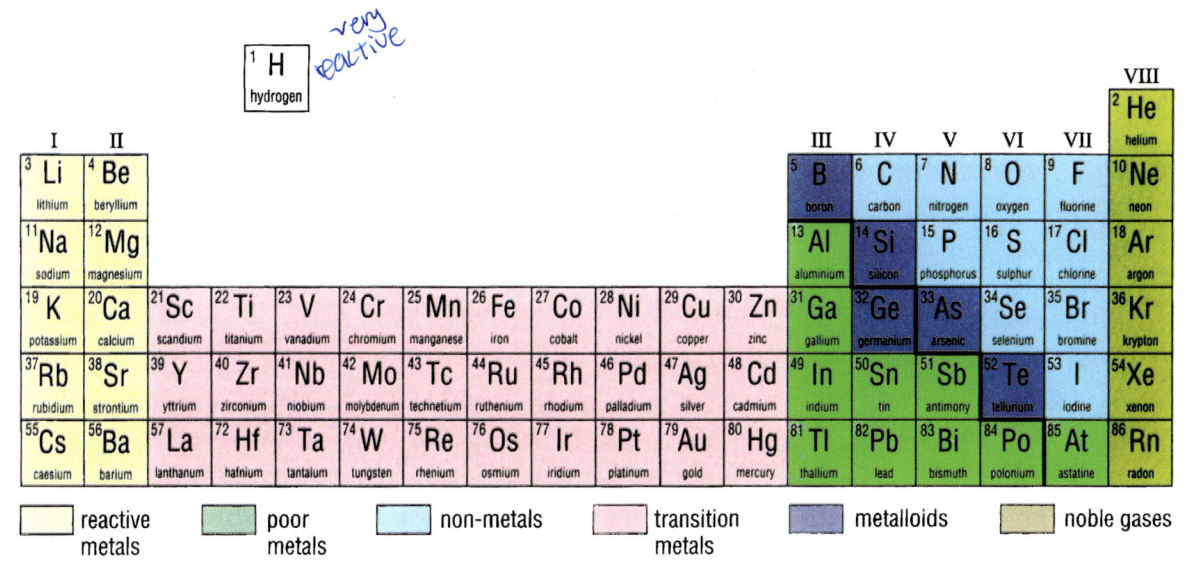

Figure 15.2 Part of the modern periodic table.

Groups of the periodic table

Many of the columns of elements in the periodic table are called groups. The elements in a group share similar properties. A trend can be seen in the properties as you go down the group.

Group I, the alkali metals

The metals in this group are not alkalis, but the oxides and hydroxides that they form are. It is this property of these compounds that gives the metals in this group their name.

Table 15.2 shows some of the physical properties of the alkali metals.

Table 15.2 Physical properties of the alkali metals.

Element	Density g/cm^3	Melting point °C	Boiling point °C
Lithium	0.53	180.6	1344
Sodium	0.97	97.9	884
Potassium	0.86	63.5	760

A closer look at the alkali metals

Lithium

Lithium's name is derived from lithis, the Greek word for stone, because it is found in many kinds of igneous rock. It is used in batteries and in compounds used as medicines to treat mental disorders.

163

THE PERIODIC TABLE

3 Which of these statements about the trends in Table 15.2 are true?
 a) The density of the metals generally:
  i) increases,
  ii) decreases down the group.
 b) The melting point of the metals generally:
  i) increases,
  ii) decreases down the group.
 c) The boiling point of the metals generally:
  i) increases,
  ii) decreases down the group.

4 Sodium is a softer metal than lithium. Describe how you think the softness of potassium compares with that of sodium.

5 Which metal in Table 15.2 has the smallest temperature range for its liquid form?

6 Which of these statements about the trends in Table 15.3 are true?
 a) The density of the metals generally:
  i) increases,
  ii) decreases down the group.
 b) The melting point of the metals generally:
  i) increases,
  ii) decreases down the group.
 c) The boiling point of the metals generally:
  i) increases,
  ii) decreases down the group.

Sodium

Metallic sodium is used in certain kinds of street lamps that give an orange glow. It is alloyed with potassium to make a material for transferring heat in a nuclear reactor. Sodium compounds such as sodium hydroxide have a wide range of uses. In the body sodium is needed by nerve cells. They use it in the transfer of electrical signals called nerve impulses.

Potassium

Potassium is used to make the fertiliser potassium nitrate. In the body it is used for the control of the water content of the blood and is used with sodium in sending electrical signals by nerve cells (see Figure 15.3).

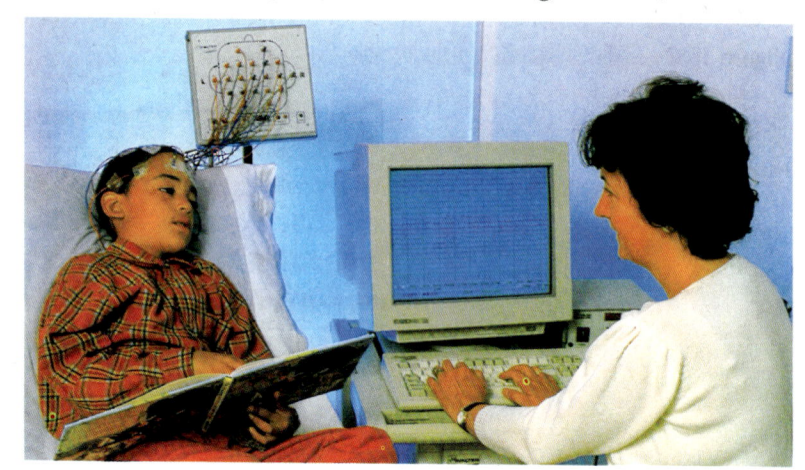

Figure 15.3 Measuring brain waves – nerve impulses are due to the movement of sodium and potassium ions in brain cells.

Group II, the alkaline earth metals

These metals are not alkalis but their oxides and hydroxides dissolve slightly in water to make alkaline solutions. Table 15.3 shows some of the physical properties of these metals.

Table 15.3 Physical properties of the alkaline earth metals.

Element	Density g/cm³	Melting point °C	Boiling point °C
Beryllium	1.85	1289	2476
Magnesium	1.74	649	1097
Calcium	1.53	840	1493

THE PERIODIC TABLE

A closer look at the alkaline earth metals

Beryllium
Beryllium combines with aluminium, silicon and oxygen to make a mineral called beryl. Emerald and aquamarine are two varieties of beryl which are used as gemstones in jewellery.

> **For discussion**
> How do the trends shown in Table 15.3 compare with those shown in Table 15.2?

Figure 15.4 Emerald (left) and aquamarine (right).

Beryllium is mixed with other metals to make alloys that are strong, yet light in weight. It is also used in a mechanism that controls the speed of neutron particles in a nuclear reactor.

Magnesium
Magnesium is used in fireworks to make a brilliant white light. Another important use is to mix it with other metals to make strong, lightweight alloys such as those used to make bicycle frames.

Green plants need magnesium in order to make the chlorophyll that traps the energy from sunlight in photosynthesis. Magnesium is needed in the body for the formation of healthy bones and teeth.

Calcium
Calcium's name is derived from the word calx, which is the Latin name for the substance lime. Lime is actually calcium oxide. Calcium forms many compounds with a wide range of uses, from baking powders and bleaching powders to medicines and plastics. In the human body calcium is required for the formation of healthy teeth and bones and for the contraction of muscles.

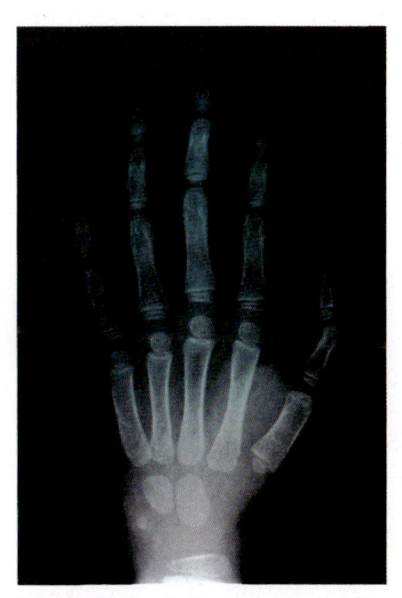

Figure 15.5 An X-ray clearly showing the bones in a hand.

THE PERIODIC TABLE

7 What trends can you see in the melting points and the boiling points of the halogens?

8 Over how many degrees Celsius is chlorine a liquid?

9 **a)** Which halogen is a liquid at a room temperature of 20 °C?
  b) Explain your answer.

Group VII, the halogens

The word halogen is a Greek word for salt former and all the elements in this group form salts readily. Table 15.4 shows some of the properties of these elements.

Table 15.4 Physical properties of the halogens.

Element	Melting point °C	Boiling point °C
Fluorine	−219.7	−188.2
Chlorine	−100.9	−34.0
Bromine	−7.3	59.1

A closer look at the halogens

Fluorine

Fluorine is a pale yellow–green poisonous gas. It is found in combination with calcium in the mineral fluorite (see Figure 15.6). This mineral glows weakly when ultraviolet light is shone on it. This property is called fluorescence. One variety of fluorite called Blue John has coloured bands and is carved into ornaments.

Figure 15.6 Fluorite glowing.

Fluorine is combined with hydrogen to make hydrogen fluoride, which dissolves glass and is used in etching glass surfaces. Sodium fluoride prevents tooth decay and is added to some drinking water supplies. Fluorine is one of the elements in chlorofluorocarbons or CFCs.

THE PERIODIC TABLE

Chlorine

Chlorine is a yellow–green poisonous gas. It is found in combination with sodium as rock salt. Chlorine is used to kill bacteria in water supply systems and is also used in the manufacture of bleach. It forms hydrochloric acid which has many uses in industry.

Bromine

Bromine is a red–brown liquid which produces a brown vapour at room temperature that has a strong smell and is poisonous. Bromine is extracted from bromide salts in sea water and is used, with silver, in photography. Silver bromide is light sensitive and is used in photographic film to record the amount of light in different parts of the image focused by the camera lens (see Figure 15.7).

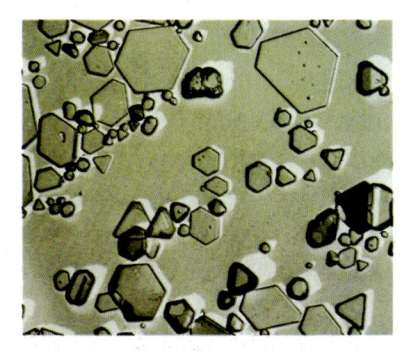

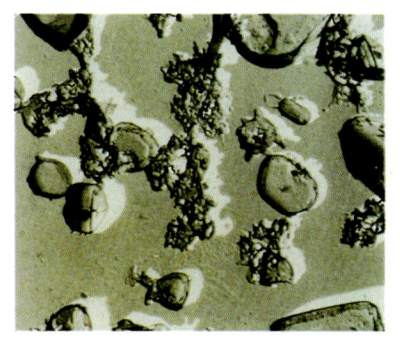

Figure 15.7 Magnified images showing silver bromide crystals on a piece of photographic film (left) and silver deposits on a developed film (right).

Noble gases

The noble gases are very unreactive.

Argon

Argon is used in light bulbs. When electricity passes through the tungsten wire in the filament the metal gets hot. If oxygen were present it would react with the hot tungsten and the filament would quickly become so thin that it would break. Argon is used instead of air containing oxygen because it does not react with the tungsten and the filament lasts longer.

It is also used in making silicon and germanium crystals for the electronics industry.

THE PERIODIC TABLE

Neon
This gas produces a red light when electricity flows through it and is used in lights for advertising displays.

Figure 15.8 Advertising displays in New York.

Helium
Helium is lighter than air and is used to lift meteorological balloons into the atmosphere. These balloons carry equipment for collecting information for weather forecasting and relay it by radio to weather stations. Helium is also mixed with oxygen to help deep-sea divers breathe underwater.

Figure 15.9 Launching a meteorological balloon.

Krypton
This is used in lamps which produce light of a high intensity, such as those used for airport landing lights and in lighthouses.

THE PERIODIC TABLE

Xenon

Xenon is used to make the bright light in a photographer's flash gun.

Figure 15.10 Press photographers, waiting for a star to appear.

Hydrogen

When you look at the periodic table (see Figure 15.2) you can see that hydrogen is placed alone. The reason for this is that its properties do not match well the properties of the other elements. Hydrogen can be considered to be unique.

The hydrogen atom is unusual in not having any neutrons. It has just one proton and electron and this makes it the lightest atom. Hydrogen is a colourless gas without any smell. It is the most common element in the Universe.

It can burn when heated with oxygen and sometimes explodes. The hydrogen 'pop' which is used in the hydrogen test is a small explosion. Hydrogen forms many compounds. It is found in acids and released when acids react with some metals. It is found in bases such as hydroxides and hydrogencarbonates. Hydrogen combines with carbon to make the hydrocarbons found in oil and is combined with nitrogen to make ammonia for use in fertilisers.

If hydrogen is mixed with air or oxygen before it is ignited, it explodes and can cause a great deal of damage. Despite the danger of explosions, cars have been developed which can run by burning hydrogen instead of petrol. The hydrogen that is needed as fuel is compressed into a tank and carefully released into the engine. The product of burning hydrogen is water vapour. There is no carbon dioxide produced as in the burning of petrol.

10 Construct a word equation for the burning of hydrogen.

11 Draw a diagram of a hydrogen atom.

For discussion

What precautions do you think should be taken if hydrogen powered vehicles were to replace petrol driven vehicles?

THE PERIODIC TABLE

♦ SUMMARY ♦

- Atoms are made from sub-atomic particles (*see page 161*).
- The number of protons in an atom is called the atomic number (*see page 162*).
- Columns in the periodic table are called groups (*see page 163*).
- Group I of the periodic table contains the alkali metals (*see page 163*).
- Group II of the periodic table contains the alkaline earth metals (*see page 164*).
- Group VII of the periodic table contains the halogens (*see page 166*).
- The noble gases (Group VIII of the periodic table) are very unreactive (*see page 167*).
- Hydrogen has unique properties (*see page 169*).

End of chapter question

Elements in the alkali metals, alkaline earth metals, noble gases and halogens are important in our lives. How accurate is this statement? Explain your answer.

Appendix

Using formulae

The symbols of most of the elements are shown in the periodic table on page 163.

The symbols are used to write formulae for elements or the compounds that the elements form. Often a number is featured in the formula. It is below the line of the letters. This number indicates the number of atoms of the element which is immediately to the left of it. For example, oxygen exists as oxygen molecules. Each one is formed from two oxygen atoms. The formula for the oxygen molecule is O_2. Other examples of molecules made from two atoms of one element are chlorine, Cl_2, and hydrogen, H_2.

A molecule of carbon dioxide has one atom of carbon and two atoms of oxygen. Its formula is CO_2.

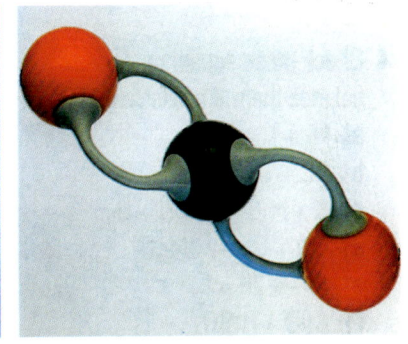

Hydrogen

Carbon dioxide

Figure A Plastic spheres can be connected together to make models of molecules.

Some molecules have three or more elements in them and there may be a different number of atoms of each element. For example, in a molecule of sulphuric acid there are two atoms of hydrogen, one atom of sulphur and four atoms of oxygen. The formula for sulphuric acid is H_2SO_4.

Calcium hydroxide is unusual in that an atom of calcium is combined with two hydroxide ions, and each hydroxide ion is made from an oxygen and a hydrogen atom. In this case brackets are used around the hydroxide ion and the number '2' is put beside them to show that two hydroxide ions are present. The formula for calcium hydroxide is $Ca(OH)_2$.

The word equation used to describe a chemical reaction can be replaced by an equation using the formulae of the compounds involved. This is called a symbol equation. It allows you to write down information about the reaction more quickly. It also

1 Write the formula for:
   a) sodium hydroxide (it has one atom each of sodium, oxygen and hydrogen),
   b) sodium nitrate (it has one atom of sodium and nitrogen and three atoms of oxygen),
   c) sodium sulphate (it has two atoms of sodium, one atom of sulphur and four atoms of oxygen).
2 Sometimes the name of a compound gives a clue to its formula. What do you think is the formula of:
   a) sulphur dioxide,
   b) carbon monoxide?
3 What does a symbol equation tell you that a word equation does not?

APPENDIX

allows more information to be given about how the atoms of the elements combine. For example, the reaction between calcium and chlorine can be written as:

$$\text{calcium} + \text{chlorine} \rightarrow \text{calcium chloride}$$
$$Ca + Cl_2 \rightarrow CaCl_2$$

Care must be taken when substituting formulae for words as the number of atoms of each element on one side of the equation must be the same as the number on the other side. To make the numbers balance, other numbers may have to be added in front of one or more of the formulae for the reactants and products. For example:

$$\text{calcium} + \text{oxygen} \rightarrow \text{calcium oxide}$$
$$Ca + O_2 \rightarrow CaO$$

In this form there are two atoms of oxygen on the left and only one on the right. The equation is balanced by adding a '2' in front of the CaO to balance the oxygen atoms and by adding a '2' in front of the Ca to balance the calcium atoms. The balanced equation is:

$$2Ca + O_2 \rightarrow 2CaO$$

When balancing equations, the formulae of the reactants and products must not be altered. For example:

$$\text{sodium} + \text{chlorine} \rightarrow \text{sodium chloride}$$
$$Na + Cl_2 \rightarrow NaCl$$

The equation is not balanced and although it could be balanced by making $NaCl_2$, this compound is not formed and the equation would be incorrect.

The equation can only be balanced by making it:

$$2Na + Cl_2 \rightarrow 2NaCl$$

4 Check these equations and balance them if necessary:
   a) $H_2 + I_2 \rightarrow HI$,
   b) $2C + O_2 \rightarrow 2CO$,
   c) $K + H_2O \rightarrow KOH + H_2$,
   d) $Mg + O_2 \rightarrow 2MgO$,
   e) $KI \rightarrow 2K + I_2$,
   f) $CuO + H_2SO_4 \rightarrow CuSO_4 + H_2O$,
   g) $H_2O_2 \rightarrow H_2O + O_2$.

Figure B By studying reactions carefully, the structure of large molecules can be discovered.

APPENDIX

5 When solid zinc oxide is placed in an aqueous solution of hydrochloric acid, zinc chloride is produced which dissolves in the water. Water is also produced.
  a) Write the word equation for this reaction.
  b) Write the equation for this reaction using the formulae in this list – HCl, $ZnCl_2$, H_2O, ZnO.
  c) Balance the equation.
  d) Write in the state symbols.

State symbols

The chemicals taking part in the reaction and the products that they form may be in different states of matter. These states can be represented by symbols in the equation. In addition to (s) for solid, (l) for liquid and (g) for gas there is a fourth symbol. It is (aq) and shows that the chemical is in an aqueous solution, which means that it is dissolved in water. The symbols are added after the formula for each chemical. For example:

calcium + hydrochloric → calcium + water + carbon
carbonate   acid   chloride   dioxide

$$CaCO_3(s) + 2HCl(aq) \rightarrow CaCl_2(aq) + H_2O(l) + CO_2(g)$$

Glossary

Some of the words in this glossary are blue. These are words which feature in the wordlists of the Cambridge Checkpoint Science (Chemistry) Scheme of Work.

A

absorbent The property of a substance to take in and hold another substance. For example, a sponge is absorbent because it can take in water.

acid A substance with a pH less than 7.0 that reacts with metals to produce hydrogen.

acidic A substance which has acidic properties.

acid rain Rain produced by the reaction of sulphur dioxide and oxides of nitrogen with water in clouds. It has a pH of less than 5.

alchemy The ancient study of chemical reactions to produce gold from less expensive metals, or to produce a chemical that would extend life.

alkali A base that is soluble in water and makes an alkaline solution.

alkali metal A soft, reactive metal with a low density and low melting point.

alkaline A condition of a liquid in which the pH is greater than 7.

allotrope One of two or more forms in which an element can exist. For example, carbon can exist as diamond or graphite.

alloy A mixture of two or more metals, or of a metal such as iron with a non-metal such as carbon.

atom A particle of an element that can take part in a chemical reaction. It contains a central nucleus which is surrounded by electrons.

atomic (proton) number The number of protons in the nucleus of an atom.

B

base A substance that can take part in a chemical reaction with an acid, forming a salt and water.

boiling A process in which a liquid turns to a vapour at the liquid's boiling point.

boiling point The highest temperature to which a liquid can be heated before the liquid turns into a gas.

brittle A hard substance that breaks easily.

C

carbonate A metal salt which contains the metal and carbon and oxygen.

catalyst A substance that speeds up a chemical reaction without being changed itself or used up in the reaction.

cell A device which contains chemicals that react and produce a current of electricity in a closed circuit.

centrifuge A machine that separates substances of different densities in a mixture by spinning them in test-tubes.

chemical change The change which takes place when the atoms in a substance combine with atoms in other substances to make new substances.

chromatography A process in which substances dissolved in a liquid are separated from each other by allowing the liquid to flow through porous paper.

combustion A chemical reaction in which a substance combines with oxygen quickly and heat is given out in the process. If a flame is produced, burning is said to take place.

compound A substance made from the atoms of two or more elements that have joined together by taking part in a chemical reaction.

concentration The amount of solute dissolved in a solvent.

condensation A process in which a gas cools and changes into a liquid.

condensing The process in which a substance changes from a gas to a liquid when it cools.

corrosion The breaking down of the solid structure of a substance.

crystal A substance made from an orderly arrangement of atoms or molecules that produces flat surfaces, arranged at certain angles to each other.

crystallisation A process in which crystals are formed from a liquid or a gas.

D

decant A process of separating a liquid from its sediment by pouring the liquid off the sediment.

decomposition A chemical reaction breaking down a substance into simpler substances.

density The mass of a substance that is found in a certain volume. A high-density substance has a volume which contains a large mass of the substance. A low-density substance has a volume which contains a small mass of the substance.

GLOSSARY

diffusion A process in which the particles in two gases or two liquids, or the particles of a solute in a solvent, mix on their own without being stirred.

displacement reaction A reaction in which a metal in a salt is replaced by another metal.

dissolving The process in which a solute mixes with a solvent to form a solution.

distil/distillation A process of separating a solute from a solvent by heating the solution they make, until the solvent turns into a gas and is condensed and collected separately without the solute.

distillate A liquid produced by distillation.

ductility The property of a substance which allows it to be pulled into a wire.

E

electrolysis The process in which a chemical decomposition occurs due to the passage of electricity through an electrolyte.

electrolyte A solution or molten solid through which a current of electricity can pass.

electron A tiny particle in an atom which moves round the nucleus. It has a negative electric charge.

electronic shell (orbit) The region around the nucleus of an atom where one or more electrons can be found.

element A substance made of one type of atom. It cannot be split up by chemical reactions into simpler substances.

endothermic A chemical reaction in which heat energy is taken in.

enzyme A chemical made by a living thing that is used to speed up chemical reactions in life processes such as digestion.

evaporating/evaporation A process in which a liquid turns into a gas without boiling.

exothermic A chemical reaction in which heat energy is released.

F

filter/filtration A process of separation of solid particles from a liquid by passing the liquid through paper with small holes in it.

fractional distillation The separation of liquids with different boiling points in a mixture by distillation.

freezing The process in which a substance changes from a liquid to a solid when its temperature is lowered.

freezing point The temperature at which a liquid changes into a solid (*see* melting point – it is the same).

G

gas A substance with a volume that changes to fill any container into which it is poured.

global warming The raising of the temperature of the atmosphere due to the greenhouse effect.

greenhouse effect The trapping in the atmosphere of the Sun's heat that is reflected from the Earth's surface.

H

hydrocarbon A compound made from hydrogen and carbon only.

hydroxide A substance containing a metal, oxygen and hydrogen.

I

igneous rock Rock formed by the cooling of magma inside the Earth's crust or lava on the surface of the crust.

immiscible A property of a liquid that does not allow it to mix with another liquid.

incandescence The glowing of a substance, due to the amount of heat that it has received.

indicator A substance which shows the pH of another substance when it is mixed with it.

irreversible reaction A chemical reaction which cannot be reversed. The products cannot take part in a reaction to directly produce the reactants again.

isotope An atom of an element that has a different number of neutrons in its nucleus to other atoms of the element.

L

liquid A substance with a definite volume that flows and takes up the shape of any container into which it is poured.

M

malleable/malleability The property of a substance which allows it to be shaped by hammering or pressing without it breaking.

mass The amount of matter in a substance. It is measured in units such as grams and kilograms.

melting The process in which a substance changes from a solid to a liquid when its temperature is raised.

melting point The temperature at which a solid turns into a liquid (*see* freezing point – it is the same).

meniscus The curved surface of a liquid where the liquid touches the surface of its container.

GLOSSARY

metal A member of a group of elements which are shiny, good conductors of heat and electricity and displace hydrogen from dilute acids.

metamorphic rock Rock formed by the effect of heat and pressure on igneous or sedimentary rock.

mineral A substance that has formed from an element or compound in the Earth and exists separately, or with other minerals to form rocks.

miscible A property of a liquid that allows it to mix freely with another liquid.

mixture An amount of matter made from two or more substances. Each substance is spread out through the other substance or substances.

molecule A group of atoms joined together that may be identical, in the molecules of an element, or different, in the molecules of a compound.

N

neutral A substance which is neither an acid nor an alkali.

neutralisation A reaction between an acid and a base in which the products (salt and water) do not have the properties of the reactants.

neutron A particle in the nucleus of an atom that has no electrical charge.

non-metal A member of a group of elements that are not shiny and do not conduct heat or electricity, or displace hydrogen from dilute acids.

nucleus The central part of an atom, which contains particles called protons and neutrons.

O

ore A rocky material that is rich in a mineral from which a metal can be extracted.

oxidation The process in which oxygen is added to a substance. The term is also used to describe the process in which hydrogen is removed from a substance.

oxide A substance formed from an element and oxygen.

P

particle A very small piece of matter.

periodic table The arrangement of the elements in order of their atomic number that allows elements with similar properties to be grouped together.

pH scale A scale on which the acidity or alkalinity of a substance is measured.

physical change The change that takes place when a substance changes from one state of matter, such as a solid, to another state of matter, such as a liquid.

precipitate Particles of a solid that form in a liquid or a gas as a result of a chemical reaction.

precipitation The process of forming a precipitate in a liquid or a gas.

pressure A measurement of a force that is acting over a certain area.

products The substances that are produced when a chemical reaction takes place.

property Something that a substance possesses. For example, if a substance bends easily it has the property of flexibility.

proton A particle in the nucleus of an atom that has a positive electrical charge.

R

reactants The substances that take part in a chemical reaction.

reactivity series The arrangement of metals in order of their reactivity with oxygen, water and acids, starting with the most reactive metal.

reversible reaction A chemical reaction which can be reversed. The products of the reaction become the reactants of the reverse reaction.

rusting The process in which iron and steel reacts with water and oxygen to form brown flakes of rust.

S

salt A compound that is formed when an acid reacts with a substance such as a metal, or when an acid reacts with a base.

sediment A collection of solid particles that settle out from a mixture of a solid and a liquid.

sedimentary rock Rock formed by particles settling out of suspensions in lakes and seas.

solid A substance that has a definite shape and volume.

soluble A property of a substance that allows it to dissolve in a solvent.

solute A substance that can dissolve in a solvent.

solution A liquid that is made from a solute and a solvent.

solvent A liquid in which a solute can dissolve.

sublimation A process in which a solid turns into a gas, or a gas turns into a solid. There is no liquid stage in this process.

sulphate A metal salt made when a metal reacts with sulphuric acid.

suspension A collection of tiny, solid particles that are spread out through a liquid or a gas.

GLOSSARY

synthesis A chemical reaction in which a substance is made from other substances.

T

temperature A measure of the hotness or coldness of a substance.

thermal decomposition The process in which heat makes a compound break down into other substances.

transparent The property of a material which allows light to pass through it without the light being scattered.

U

universal indicator An indicator which can be used to test a substance to see if it is acidic or alkaline.

V

volume The space occupied by a certain amount of matter.

W

water cycle The movement of water from the atmosphere to the land and oceans, and then back to the atmosphere again.

X

X-rays Electromagnetic waves with wavelength between ultraviolet and gamma rays.

Index

Note: page numbers in *italics* refer to entries in the Glossary.

absorbent materials 48, *174*
Acheson process 118
acid rain 27, *174*
acids 20–1, *174*
   detection of 23–4
   nitric 91–2
   reaction with metals 129–30
   reaction with non-metals 130–1
   strength of 24–5
   see also hydrochloric acid; sulphuric acid
aerosol sprays 31, 32
air 32, 63
   composition of 117
   fractional distillation 105
alchemists' symbols 78, 79
alchemy 2–4, 19, *174*
alembic 3
alkali metals (group I) 163–4, *174*
alkaline earth metals (group II) 164–5
alkalis 22–3, 127, *174*
   detection of 23–4
   strength of 24–5
allotropes 117, *174*
alloys 111, *174*
   of beryllium 165
   of bronze and brass 113
   of magnesium 165
aluminium, oxidation reaction 135
ammonia 28
   production of 92
ammonium nitrate 92
ammonium sulphate 89
animals
   acids 20
   body temperature 151
anode 102
antibumping granules 72
antioxidants 133–4
apparatus 10–12
aquamarine 165
aqueous solution 173
argon 167
ascorbic acid (vitamin C) 20
atmospheric pressure 59–60
atomic number 162, *174*
atoms 80–1, *174*
   electron shells 161–2
   mass 81

bad laboratory practice 15–16
balancing equations 172
balloons, use of helium 168
barium sulphate, use in X-rays 90, 165
bases 22–3, *174*
batteries, use of sulphuric acid 89
bee stings, neutralisation 25
beryl 165
beryllium 165
beryllium atom 80

Berzelius, Jöns Jakob 79
bicarbonate of soda *see* sodium
      hydrogencarbonate
bicycle tyres 32
biological catalysts (enzymes) 158–9
bitumen 106
blast furnace 114–15
Blue John 166
body
   use of calcium 165
   use of magnesium 165
   use of sodium and potassium 164
body temperature of animals 151
boiling 36–7, *174*
   particle theory 57
boiling point *174*
   effect of altering pressure 58–60
   of solutions 101
   use in testing for purity 61
bones, need for calcium and magnesium 165
Bosch, Carl 92
Boyle, Robert 4
brain waves 164
brake fluid 31
brass 113
bricks, absorbency 48
brittle materials 48–9, *174*
bromine 78, 166, 167
   origin of name 80
bronze 113, 115
   patina formation 136
Buchner funnel 69
buckminsterfullerene 117
Bunsen burner 10, 11, 41–2
Bunsen, Robert 11
burettes 6, 7
burning 40–3, 144
   burning candle investigation 144–5
   incomplete combustion 146

calcium 165
   flame colour 83
   reaction with water 128–9
calcium carbonate 96
   reaction with hydrochloric acid 142, 153
   reaction with nitric acid 92
   reaction with sulphuric acid 91
calcium chloride 142, 172
calcium hydroxide (slaked lime) 95
calcium hydroxide molecule 171
calcium hydroxide solution (lime water) 93
calcium oxide (lime, quicklime) 28, 95, 165, 172
   limelight 96
   use on acid soil 26
calcium silicate (slag) 115
calcium sulphate 89
   gypsum 90
calorimeter 147–8
candles, investigation of burning 145

carbon 117
   charcoal 119–20
   combustion 93
   diamond 118–19
   formation in stars 1
   graphite 53, 118
   reaction with acids 130
   use for electrodes 102
carbonates 96–7, *174*
carbon atom, structure 162
carbon dioxide 83, 128
   production in respiration 148
   solid (dry ice) 38–9
   testing for 93
   use in photosynthesis 149
carbon dioxide molecule 171
carbon monoxide 146
car engines, catalytic converters 157–8
cast iron 116–17
catalysts 156, 159, *174*
   biological 158–9
   use in industry 157
cathode 102
catalytic converters 157–8
caustic soda *see* sodium hydroxide
cell *174*
Celsius scale 9
centrifuge 69, *174*
chalcopyrite 112
changes of state
   particle theory 55–7
   and pressure 59
charcoal 117, 119–20
   combustion 93
chemical change *174*
chemical plants 18
chemical reactions
   description by equations 18
   first use by humans 2
chemical symbols 78–9
chemistry, what it is 6
China, alchemists 2–3
chip pan fires 43
chlorides of metals
   preparation of 87–8
      sodium chloride 85–6
chlorine 78, 166, 167
   origin of name 80
chlorine molecule 171
chlorofluorocarbons (CFCs) 166
chlorophyll, need for magnesium 165
chromatography 71, *174*
chromium plating of steel 135
circuits, for testing conductivity 51, 52
citric acid 20
   reaction with sodium hydrogencarbonate 149
climate change 35
clinical thermometers 9
clouds 38

INDEX

coal 126
coal dust, as cause of explosions 133, 154
coke 120
coloured hydroxides 95–6
combustion 40–3, 144, *174*
    of carbon 93
    of metals 94
    see also burning
complete combustion 145
compounds 78, 83, 99–100, *174*
    proportions of elements 83
    separation 101–3
    solutions 100–1
    synthesis reactions 100
compressed gases 31
concentration *174*
    effect on rate of reaction 154, 159
condensation 37, *174*
    particle theory 57
conductors
    of electricity 51–3
    of heat 51
cooking, endothermic reactions 150
copper 109–10
    extraction from ore 110, 112, 125
    flame colour 83
    properties and uses 113
    reactivity 139
        with hydrochloric acid 130
        with oxygen 137
        with water 129
    use by Egyptians 2
    verdigris 136
copper carbonate 97
    thermal decomposition 43
copper oxide, reaction with sulphuric acid 90, 142
copper sulphate 90, 91
    preparation of 142
corrosion *174*
    fast and slow reactions 133–4
    of metals 134–6
corrosive substances, warning sign 16
cost-benefit analysis 5
crystallisation 70, *174*
cubic centimetres (cm^3) 6

damp-proof course 48
DDT 122
decanting 68, *174*
decomposition *174*
density 29, *174*
    comparison of metals and non-metals 109
detergent manufacture use of sulphuric acid 89
diamond 117, 118–19
diesel oil (gas oil) 106
diffusion 60–1, *175*
digestion 150–1
digestive enzymes 158
discovery of elements 75–7
displacement reactions 138, *175*
dissolving 58, *175*
distillation 72, *175*
    with Liebig condenser 73
distilled water 70
divers, use of helium 168
dry ice 38–9
ductility *175*

Egyptians, use of copper 2
einsteinium 80
electrical charges of sub-atomic particles 80
electrical conductors 51–3
electrodes 102
electrolysis 102–3, *175*
electrolytes *175*
electronics industry, use of argon 167
electrons 80–1, *175*
electron shells 161–2, *175*
electroplated nickel silver (EPNS) 112
elements *175*
    ancient Greek 34
    chemical symbols 78–9
    discovery of 75–7
    flame tests 82–3
    lightest 81–2
    names of 79–80
    periodic table 82, 163–70
    properties of 78
    separation from compounds 102–3
elixir of life 2
emerald 165
emulsions 63
endothermic reactions 144, 149–51, *175*
energy release, measurement of 147–8
energy sources 126
enzymes 158–9, *175*
Epsom salts (magnesium sulphate) 90
equations 18
etching, use of hydrogen fluoride 166
ethanoic acid 21
ethanol
    properties 52
    separation from water 104
    as solvent 66
evaporation 36, *175*
    as method of separation 70
    particle theory 56
exothermic reactions 144, *175*
    burning 144–6
    calorimeters 147–8
    respiration 148
explosions 133, 152
explosive substances, warning sign 16

fast corrosive reactions 133
'father of chemistry' 3
fats, oxidation of 133–4
fertilisers
    ammonium nitrate 92
    manufacture using of sulphuric acid 89
filters, use of charcoal 120
filtrate 69
filtration 68–9, *175*
fire, first use by humans 2, 19
fire extinguishers 26
fire-fighting 42–3
fire triangle 42
fireworks 83
    invention by Chinese 2
first humans 1–2
fixed proportions 83
flammable substances, warning sign 16
flame tests 82–3
flash photography, use of xenon 168
flasks 10

flexible materials 49
flotation cell 67
fluorine 166
fluorite 166
foams 63
fog 63
formulae 171–3
fossil fuels 126
fractional distillation 104–5, *175*
    of oil 105–7
fractionating tower 106
Frasch, Herman, Frasch process 120–1
freezing 34–6, *175*
    particle theory 56
freezing points *175*
    of solutions 100
fuel efficiency 146
fuel oil 106
fuels
    hydrogen 169
    measuring energy release 147–8
fungicides 121

gaseous elements 78
gases 29, 30, *175*
    diffusion 60
    measurement of volume 7
    in mixtures 63
    movement of particles 55
    pressure 58
    properties of 31
gas fires, need for ventilation 146
gas masks, use of charcoal 120
gasoline 106
Gerber (Jabir Ibn Haiyan) 3
glass
    properties 46, 48–9, 50
    recycling 125
    use for laboratory apparatus 10
global warming *175*
gloss paint 66
glucose
    production in photosynthesis 149
    use in respiration 148
gold 109–10, 125
    formation in supernovas 1
    reactivity 139
        with oxygen 137
graphite 117, 118
    properties 53
Greek philosophers, ideas about matter 33–4
greenhouse effect 35, *175*
ground-glass joints 16, 17
groups of periodic table 163
    alkali metals (group I) 163–4
    alkaline earth metals (group II) 164–5
    halogens (group VII) 166–7
    noble gases (group VIII) 167–9
gunpowder 2
gypsum (calcium sulphate) 90

Haber, Fritz 92
habitat destruction 122, 123
haematite 111, 114
halite (rock salt) 86
halogens (group VII) 165–6
hardness, comparison of metals and non-metals 109

179

INDEX

harmful substances, warning sign 16
heat, effect on particles 55–6
heat conductors 51
heat energy, recycling of 125–6
helium 168
    formation in stars 1
Hittites, use of iron 2, 115–16
household cleaners, use of alkalis 23
hydrangeas 23
hydrocarbons 105, *175*
    burning 145
    fractional distillation 105–7
hydrochloric acid 20, 21
    in production of metal chlorides 85–8
    reaction with calcium carbonate (marble chips) 153
    reaction with metals 129–30
        magnesium 153
        zinc 141
    reaction with potassium hydroxide 143
    reaction with sodium thiosulphate solution 155
hydrogen 169
    in stars 1
    testing for 94
hydrogen fluoride 166
hydrogen ions, in acids 24
hydrogen molecule 171
hydrogen peroxide 156–7
hydroxide ions in alkalis 24
hydroxides 94–6, *175*

ice, effect of pressure 59
igneous rock *175*
immiscible *175*
immiscible liquids, separation 73–4
incandescence *175*
incomplete combustion 146
India, ancient use of wrought iron 116
indicators 24–5, *175*
indigestion tablets 26
ink, chromatography 71
insulators (heat) 51
iodine
    properties 53
    sublimation 39
ions 81
iron
    chemical symbol 79
    extraction from ore 110, 114–15
    formation in stars 1
    history of extraction and use 2, 115–16
    as non-renewable material 123
    properties 99
    reaction with hydrochloric acid 130
    reaction with steam 129
    reactivity 137
    rusting 134–5
    uses 116–17
iron sulphide 83, 99–100
irreversible reactions *175*
    burning 41
irritant substances, warning sign 16
isotopes 162, *175*
issues 5, 19

Jabir Ibn Haiyan (Gerber) 3

Kando, mining issue 5
kerosene 106
kimberlite 118
kinetic theory, and rates of reaction 159
krypton 168

laboratory rules 13–14
lacquer use by Chinese 3
lactic acid 20
laminated glass 49
lamps, krypton 168
Lavoisier 79
leaching 27
lead, reaction with hydrochloric acid 129
lead atom, structure 162
lead bromide, electrolysis 103
lead iodide, preparation of 143
lead nitrate, reaction with potassium iodide 143
Liebig condenser 12, 13, 73
light bulbs, use of argon 167
lightning, production of nitrogen oxides 91
lime 28, 165
    use on acid soil 26
    see also calcium oxide
lime kiln 150
limelight 96
limestone 48, 96
    decomposition 150
    use in blast furnace 114–15
lime water 93
liquid elements 78
liquid mixtures, separation 104–7
liquids 29, 30, *175*
    arrangement of particles 55
    diffusion 60–1
    measurement of mass 8
    measurement of volume 6–7
    in mixtures 62, 63
    pressure 58
    properties of 31
    testing for conduction of electricity 52
lithium 163
litmus 23
liver enzymes 158
lubricating oil 106
luminous flame, Bunsen burner 42

magnesium 164–5
    burning 43, 53
    combustion 94
    reaction with hydrochloric acid 130, 153
    reactivity 139
        with water 129
magnesium oxide 83, 128
    reaction with nitric acid 92
magnesium sulphate (Epsom salts) 85, 90
magnetic metals 109
magnetic separators 66–7, 125
malachite 97
malleable materials 50, *175*
manganese dioxide, use as catalyst 156–7
marble chips, reaction with hydrochloric acid 142, 153
mass *175*
    of atoms 81
    and changes of state 40
    measurement of 7–8

mass of reactants, changes in 153
materials 45
    properties 46–53
matter
    chemical changes 40–3
    early ideas about 33–4
    physical changes 34–40
    properties of 30–2
    states of 29–30
        plasma 81
measuring
    mass of solid or liquid 7–8
    temperature 8–9
    volume of gas 7
    volume of liquid 6–7
measuring cylinders 6
melting 34–6, *175*
    particle theory 55–6
melting point *175*
    comparison of metals and non-metals 109
    use in testing for purity 61
meniscus 6, 7, *175*
mercury 78
    meniscus 7
    properties 52
Mesopotamia, alchemy 2
metal chlorides
    preparation of 87–8
    sodium chloride 85–6
metals *176*
    alloys 111
    comparisons with non-metals 109, 110
    extraction from ore 110, 139
        copper 112
        iron 114–15
        silver 111
    flame tests 82–3
    properties 46, 50, 53
        conduction of electricity 51
    reaction with acid 129–30
    reaction with non-metals 127
    reaction with water 128–9
    recycling 125
metamorphic rock *176*
meteorites, iron content 115
methane
    burning 145
    production in decomposing waste 122
methanoic acid 20, 25
milk 63
millilitres (ml) 6
mineral *176*
mineral acids 21
    see also acids
mining 122–3
miscibility, effect of temperature 65
miscible *176*
mist 63
mixtures 62–3, *176*
    of elements 99
    proportions of components 84
    separation of 66–74
        liquids 104–7
        solutions 64–6
molecules 171, *176*
Moon, mining of 126
Muslim alchemists 3

INDEX

naphtha 106
native metals 110
    iron as 115
natural gas 105, 106, 126
    burning 145
neon 168
nerve cells, use of sodium and potassium 164
nettle stings, neutralisation 25
neutralisation reactions 22, 25–6, *176*
    preparation of sodium chloride 85–6
neutrons 80, *176*
    number in nucleus 162
nitrates 91, 92
nitric acid 21, 91–2
    reaction with carbon 130
    reaction with sulphur 131
nitric oxide production, use of catalyst 157
nitrogen 78
nitrogen dioxide 130
nitrogen oxides
    and acid rain 27
    production by lightning 91
noble gases 167–9
non-luminous flame, Bunsen burner 42
non-metals *176*
    comparisons with metals 109, 110
    properties 51, 53
    reaction with acid 130–1
    reaction with metals 127
    reaction with oxygen 127–8
    reaction with water 129
non-renewable materials 123–4
nuclear reactors
    use of beryllium 165
    use of sodium and potassium 164
nucleus of atom 80, *176*

oil 126
    fractional distillation 105–7
    use in engines 31
oil fires 43
oils, oxidation of 133–4
oleum 89
opaque materials 50
open cast mining 122–3
ores *176*
    extraction of metals 67, 110, 139
        copper 112
        iron 114–15
        silver 111
    extraction of sulphur 120–1
organic acids 21
oxidation reactions 133, *176*
    of metals 135–6
    rusting 134–5
oxides 127–8, *176*
oxidising substances, warning sign 16
oxygen 78
    formation in stars 1
    production in photosynthesis 149
    reaction with metals 127
    reaction with non-metals 127–8
    reactivity of metals 137
    requirement for burning 145
    solubility in water 66
    testing for 93
    use in respiration 148
oxygen molecule 171

paper
    invention by Chinese 3
    recycling 125
particle *176*
particle size, effect on rate of reaction 154, 159
particle theory of matter 55
    changes of state 55–8
    diffusion 60
    pressure 58–60
    and rates of reaction 159
pencil 'lead' 118
periodic table 82, 163, *176*
    group I (alkali metals) 163–4
    group II (alkaline earth metals) 164–5
    group VII (halogens) 166–7
    group VIII (noble gases) 167–9
    hydrogen 169
pesticide pollution 122
petrochemicals industry 105–7
petrol fires 43
pH 24–5, *176*
    effect on enzyme-catalysed reactions 158
philosopher's stone 2
photography
    use of silver bromide 167
    use of xenon 169
photosynthesis 149
physical change *176*
pig iron 115
plants
    acids 20
    use of magnesium 165
plasma 81
plaster of Paris 90
plastic
    manufacture from milk and vinegar 4
    properties 46
plasticine, properties 50
platinum
    use as catalyst 157
    use for electrodes 102
pollution, of land 122
potassium 164
    chemical symbol 79
    flame colour 83
    reaction with water 129
potassium chloride, preparation of 142–3
potassium hydroxide 22
    reaction with hydrochloric acid 143
potassium iodide, reaction with lead nitrate 143
potassium nitrate 85
pottery
    first use 2
    properties 46, 50
power stations, sulphur dioxide production 88
precipitation (of solids) 101–2, *176*
precipitation (in water cycle) 38
pressure 58, *176*
    atmospheric 59–60
    and changes of state 59
products 18, *176*
    change in volume 153–4
propanone, as solvent 66
properties *176*
    of materials 46–53
    of matter 30–2
    of metals and non-metals 110
proportions

    in compounds 83
    in mixtures 84
protons 80–1, *176*
    and atomic number 162
purity, testing for 61, 72
Pyrex (borosilicate glass) 10

quicklime *see* calcium oxide

radioactive materials
    pollution of land 122
    warning sign for 16
rainforests, open cast mining 122–3
rancid food 133
rates of reaction 133–4, 152
    catalysts 156–9
    change in mass of reactants 153
    change in volume of product 153–4
    effect of concentration 154
    effect of particle size 154
    effect of temperature 155–6
    kinetic theory 159
reactants 18, *176*
    changes in mass 153
reactions
    displacement 138
    endothermic 144, 149–51
    exothermic 144–8
    fast 133
    irreversible 41
    rates of 152–6
        catalysts 156–9
    reversible 18
    slow 133–4, 152
    synthesis 100
reactivity 137
    displacement reactions 138
    and metal extraction 139
    of metals with oxygen 137
reactivity series 139, *176*
recycling 124–6
renewable materials 123
research laboratories 17
reserves of raw materials 124
resources of raw materials 124
respiration 148
reversible reactions 18, *176*
rhodium, use as catalyst 157
rocks, absorbent 48
rock salt (halite) 86, 167
rubber, vulcanised 121
rules of laboratory 13–14
rusting 134–5, *176*

safety in laboratory 13–14
salad dressing 73
salt (sodium chloride) 78
    effect on rusting 135
    preparation of 85–6
    use on icy roads 100–1
    see also sodium chloride
salt pans 86, 87
saltpetre 2
salts 85, *176*
    preparation of 141–3
sandstone 48
saturated solutions 65
sea water, saltiness of 86

181

INDEX

sediment *176*
sedimentary rock *176*
separating funnel 10, 74
separation
    of compounds 101–3
    of immiscible liquids 73–4
    of miscible liquids 104–7
    of recycled materials 125
    of solid/liquid mixtures 68–9
    of solid/solid mixtures 66–7
    of solutes from solute/solvent mixtures 70–1
    of solvents from solute/solvent mixtures 72–3
shape, relationship to properties of materials 47
sherbet 149
sieves 66, 68
silver
    chemical symbol 79
    extraction and uses 111–12, 125
    formation in supernovas 1
    reactivity 139
    tarnishing 136
silver bromide, use in photography 167
silver chloride, precipitation 102
slag (calcium silicate) 115
slaked lime *see* calcium hydroxide
slow reactions 133–4, 152
smoke 63
soap 23
soda-acid fire extinguisher 26
sodium 78, 164
    chemical symbol 79
    combustion 94
    flame colour 83
    properties 109
    reactivity 137, 139
        with oxygen 137
        with water 129
sodium chloride 171
    electrolysis 103
    preparation of 85–6
    see also salt
sodium fluoride, addition to water 166
sodium hydrogencarbonate (sodium bicarbonate) 97
    reaction with citric acid 149
sodium hydroxide 22, 94–5
    reaction with hydrochloric acid 85–6
    reaction with nitric acid 92
sodium sulphate 90
sodium thiosulphate solution, reaction with hydrochloric acid 155
soil 62
    acidity 26
soil sieve 66
solid/liquid mixtures, separation 68–9
solids 29–30, *176*
    arrangement of particles 55
    in mixtures 62–3
    pressure 58
    properties of 30–1
solid/solid mixtures, separation 66–7
solubility 65–6
soluble and insoluble substances 64

solutes 64, *176*
    separation from solute/solvent mixtures 70–1
solutions 62, 64–5, 100–1, *176*
    conduction of electricity 53
    formation of 58
solvents 58, 66
    separation from solute/solvent mixtures 72–3
stainless steel 135
standard pressure 59
stars
    as chemical factories 1
    plasma 81
states of matter 29–30
    plasma 81
state symbols 173
steam 37
steel 116, 117
    rusting 134
    rust prevention 135
stings, neutralisation 25
stocks of raw materials 123–4
stoves, improving efficiency 146
street lamps, use of sodium 164
strontium, origin of name 80
sub-atomic particles 80–1
sublimation 38–9, 57, *176*
suction pump, use in filtration 69
sulphates 89–91, *176*
sulphur 120
    extraction 120–1
    properties 99, 121
    reaction with acids 131
sulphur dioxide 130
    and acid rain 27, 88
    removal from waste gases 28
sulphuric acid 21, 88–9
    reaction with calcium carbonate 91
    reaction with carbon 130
    reaction with copper oxide 90, 142
    reaction with sulphur 131
    reaction with zinc 89
sulphuric acid molecule 171
sulphuric acid production, use of catalyst 157
sulphur trioxide 88–9
sulphur vapour, sublimation 39, 57
supernovas 1
superphosphate 89
surface area, effect on reaction rates 154
suspensions 62, *176*
symbol equations 171–3
symbols of elements 78–9, 171, 173
synthesis reactions 100, *177*
syringe, use to measure volume of gas 7

tare, use of 8
tarnishing of silver 136
tartaric acid 20
teaching laboratories 16
teeth, need for calcium and magnesium 165
temperature *177*
    effect on rate of reaction 155–6, 159
    effect on solubility and miscibility 65–6
    measurement of 8–9
test tubes 10

Thales 33
thermal decomposition 43, *177*
thermometers 8–9
tin, in bronze 113
titanium 89
top loading balance 7–8
toughened glass 48
toxic substances, warning sign 16
translucent materials 50
transparent materials 50, *177*
transpiration 38
triangle of fire 42
tyres, recycling of 124

universal indicator 24–5, *177*
universal solvent 66
uranium, origin of name 80
uric acid 20

vanadium oxide catalysts 88, 157
verdigris 136
vinegar, ethanoic acid 21
vitamin C (ascorbic acid) 20
volcano, model of 22
volume *177*
volume of product, change in 153–4
vulcanised rubber 121

warning signs 16
washing powders, use of enzymes 158–9
wasp stings, neutralisation 25
waste disposal 122
water
    addition of chlorine 167
    as a compound 83
    conduction of electricity 52
    electrolysis 103
    freezing point 35
    reaction with metals 128–9
    separation from ethanol 104
    solubility of oxygen 66
    testing for purity 61, 72
water cycle 38, *177*
waterproof materials 47
water vapour 38
weight of atoms 81
white spirit 66
wine, conversion to ethanoic acid 21
wood
    properties 46, 47, 49
    as renewable material 123
word equations 18
wrought iron 116

xenon 169
X-rays, barium meals 90, 165, *177*

zinc
    in brass 113
    oxidation 135
    reaction with hydrochloric acid 130
    reactivity 139
        with sulphuric acid 89
zinc chloride, preparation of 87–8, 141